English for Engineers

کلیو براسنت

انگلیسی برای مهندسان

ناشر: انتشارات نیما ـ تبریز

تعداد: ۲۰۰۰ جلد

قطع: رقعی ۱۴۴

نوبت چاپ: چاپ سوم، زمستان ۱۳۶۷

فیلم وزینگ: لیتوگرافی تصویر

چاپ: شرکت طلوع چاپ

بها ۵۶۰ ریال

English for Engineers

Clive Brasnett

Methuen Educational Ltd
LONDON · TORONTO · WELLINGTON · SYDNEY

First published by Methuen Educational Ltd 1968
11 New Fetter Lane, London EC4
Reprinted 1970
Reprinted 1971
Reprinted 1974
Reprinted 1976

Printed in Great Britain by
Fletcher & Son Ltd, Norwich

ISBN 0 423 84130 0

Contents

Preface

The passages in this book were devised during the course of three years' work with students of the Technological Institute of Damascus. It was found that although a limited amount of material on purely scientific subjects adapted into English suitable for foreign students was available, very little could be found to meet the interests and requirements of students intending to specialise in one of the applied sciences. Passages were, therefore, chosen from works dealing with various aspects of engineering and technology and rendered into an English suitable for students whose knowledge of the language was initially fairly elementary.

However, the level of difficulty, as far as the language is concerned, is carefully graded throughout the book, as a comparison of the vocabulary and structures used in the earlier and later passages will readily reveal. For this reason, although it is possible to select for study certain sections which are of particular interest to particular groups of students, since each section has a unity of its own, particularly with regard to the recurrence of certain technical terms, teachers may find that the greatest benefit is to be gained by proceeding through the passages in the order in which they appear in the book.

Words and phrases which may be expected to cause students difficulty are printed in **bold type,** and a glossary of these is provided immediately after the passage in which they have occurred. This arrangement is made for ease of reference, but these words and phrases are also indexed at the end of the book so that the student can readily find the page on which they were first explained. For the most part, explanations given are contextual. A term whose meaning may vary to some extent as it is used in different branches of science is explained according to its particular usage within the passage. However, where there is some affinity between the use of a term in a scientific context and in another context which may already be familiar to the student, as with **current,** for example, attention is drawn to this affinity.

Permission from authors and publishers to make use of printed and illustrative material is gratefully acknowledged. The title of the work adapted will be found at the end of each passage, and it is hoped that students will be enabled and encouraged to go on to read these and similar textbooks in their original form. I am also grateful to successive teaching colleagues at the Technological Institute of Damascus, Mr Abdul Hadi Abla, Mr Muhammad Mohie ed-Deen, Mr Abdullah Hassoun, and Mr Teyssir Kamleh, all of whom have given me valuable advice in the light of their experiences making use of these and similar texts in teaching English to various groups of students. Warm thanks are also due to Mr Youssef Joubeily, who was kind enough to comment on the presentation of material in certain passages from the point of view of a practising engineer, and to Miss Aida Adham, for unremitting work in the preparation of typescript throughout a series of changes and additions.

1968 **Clive Brasnett**

Communications

1. Messages by Electricity

Electricity completely changed communications. Once it was **discovered** that an electric **current** would flow along a wire, it seemed possible that it could be used for messages. As long as two places were connected by a wire, they could send electric currents to each other. They could send over hills and round corners, and they could send them with the speed of light. The question was how could electric current make words?

Real telegraphs were not possible until it was proved that electricity and magnetism were connected. A Dane discovered this. He was called Hans Oersted. He found in 1819 that he could make a needle of a compass move by putting it near a wire with a current flowing through it.

Then a Frenchman named Ampère thought about this discovery. He reasoned like this. A compass needle is a magnet. If a wire carrying a current could move it, then the wire must be a magnet too. This he found to be true. He also found that the magnetic force round the wire could be **concentrated** by **coiling** the wire. He had, in fact, made an **electro-magnet.**

A lot of people began to see how electro-magnets could be used to send messages. Two men, called Wheatstone and Cooke, laid a **telegraph line** between Euston Station and Camden Town, in London. The way they used electricity to send messages was to have five compass needles swinging in the centre of wire coils at each end of the telegraph line. (See figure 1.) Their movements showed letters of the alphabet. When Cooke, in Camden Town, sent a current along one of the wires it went through a coil in front of Wheatstone, in Euston. The magnetic force made the needles swing. Wheatstone read the letter the needles pointed to and **gradually got the message.** Then he switched the current through to Cooke by pressing a 'key' and sent a reply.

But one of the most famous people of all in communications is Samuel Morse. He **invented** the Morse code. This code is still

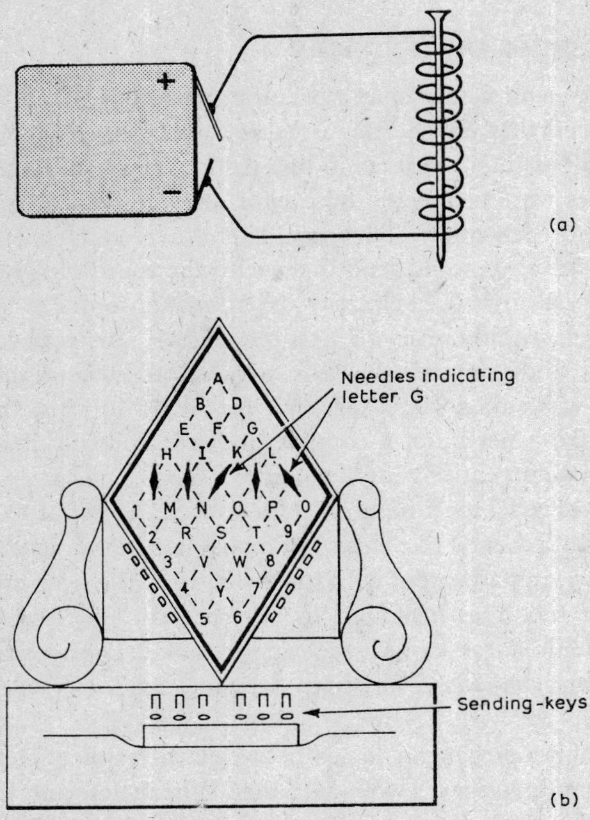

Figure 1

used in **signalling** all over the world. His idea was simple. He thought of using time, sending current along a wire for a long time or a short time. **Combining** different sets of these **'impulses'**, just using the long or short, he made a code for the whole alphabet, and the numbers from 0 to 9. For instance, E is the most commonly used letter in the alphabet, so Morse used the **simplest** signal for it – a short impulse, which is called a 'dot'. T is the next most commonly used letter. Morse used the next simplest signal – a long impulse, which is called a 'dash'. He went through the alphabet in this way, making up signals for all the letters, in 'dots' and 'dashes'. Here are some of the letters:

A . — B — . . . C — . — . D — . . E . F . . — . G — — .

Everyone knows the Morse code for S O S, which is the distress signal for ships in serious trouble. It is . . . — — — . . . It is the simplest signal to make in times of danger or disaster.

The Morse code **simplified** sending messages. It was also very fast. With practice, Morse code can be read at 30 words a minute or more. It became by far the most widely used code because it was the easiest to understand and the fastest to **transmit.** By 1871, for instance, there were 6,000 telegraph stations in the United States, all working on the Morse code system.

The telegraph meant that at last people could send messages a long way at great speed. Soon people wanted to send messages even farther.

If two places on land, say, London and Manchester, could be **connected** by wire, why could not the lands themselves – England and America, for instance – be connected by wires under the sea? This was a question quickly answered after the telegraph was proved to work. There were some problems about this, however. The main one was insulation, which means protecting the wire so that the electric current does not **leak away.** Dry air is a good insulator, but sea water is not. If a wire was laid under water, the current would all **leak away.** In any case, all the insulating substances known up to then could be damaged by sea water. But in 1847 gutta-percha was **introduced** as an insulating material. Gutta-percha is **a gummy substance** obtained from trees, like rubber, and it does not allow electricity to flow through it easily.

After several unsuccessful attempts, in 1866 Britain and America were **linked** by underwater **cable** by the famous cable-laying ship the *Great Eastern.* After that cables were laid the world over, and **nowadays** it is possible to telegraph from almost everywhere.

Adapted from *Over to You* by Roy Herbert (Brockhampton Press)

Explanation of words in bold type

discovered found out for the first time.
discovery something that is found out for the first time.
a current flows along in one direction. You can see the **current** in a river flowing towards the sea. Electric current flows along a wire when a switch is turned on.
concentrated kept together in one place.
coiling winding something round to look like a spring, which we can also call 'making a coil'.
electro-magnet a magnet made by coiling wire round a piece of metal. When an electric current flows through the wire, the metal becomes a magnet.
a telegraph line a wire, or several wires, used to send messages by electric current.
gradually got the message received the message slowly, one letter at a time.
invented to **invent** is to make something for the first time. Morse was an **inventor.** His **invention** was the Morse code.
signalling sending messages by signs, or 'signals'.
combining putting together.
impulse a flow of electric current lasting a very short time.
simplest easiest.
simplified made easy.
transmit send.
connected joined.
leak away If there is a hole in the kettle the water in it will **leak away.**
introduced first used.

a gummy substance a material that feels like gum, or thick glue.
linked joined together.
cable a tube carrying electric wires. Telegraph cables carry messages under the sea. Messages sent in this way are called 'cables'. Telegraph messages which travel between people in the same country are called 'telegrams'.
nowadays at the time we are living in now.

Exercises

A Use these words to complete the following sentences:
cables, connect, electricity, impulses, insulation, key, longer, magnets, simple, telegraph.

1 Hans Oersted discovered that . . . . and magnetism are connected.
2 A wire carrying an electric current can move a compass needle because they are both . . . .
3 Cooke and Wheatstone sent messages to each other through a . . . . line
4 Wheatstone pressed a . . . . to switch the current through to Cooke.
5 The dots and dashes used in the Morse code can be made by electric . . . . passing through a wire.
6 The impulse used to send a dash is . . . . than that used to send a dot.
7 The letters S O S are used in times of danger because they are . . . . letters to signal.
8 Telegraph lines . . . . telephone stations that are far apart.
9 . . . . is used to protect telegraph wires.
10 Messages are carried under the sea through telegraph . . . .

B Answer these questions.

1 At what speed can messages be sent by electricity?
2 How did Hans Oersted make the compass needle move?
3 How did Ampère concentrate the electric force around a wire?
4 What did Ampère make?

5 What made the needles in Cooke and Wheatstone's invention point to the letters?
6 Why is the simplest signal in the Morse code used for the letter E?
7 Write the signals used for AGE in the Morse code.
8 At what speed can Morse code messages be read?
9 Give *two* reasons why the Morse code was so widely used.
10 What was the main problem about sending messages by telegraph under the sea?
11 Give *two* reasons why telegraph cables must be insulated.
12 Why is gutta-percha a good material for insulating telegraph cables?

2. The Electric Telegraph

Modern telegraph systems are based upon the fact that electric current will flow through a wire which forms **a circuit.** The circuit can be broken to **interrupt** the flow of electricity and closed again for a short time by a key to send an electrical impulse through the circuit. Look at figure 2. Here the wire is not

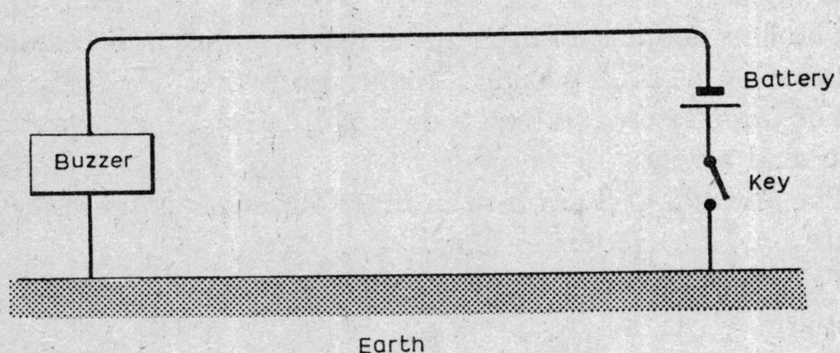

Figure 2

joined all the way round, but the circuit is complete because electricity will flow through the wire and back through the earth. This is called an 'earth return'. When the key is pressed the circuit is closed. Electricity from the battery flows through the wire and earth circuit and causes **the buzzer** to make a noise. Using the Morse code, or some other code, messages can be sent in this way.

Men wanted to **increase** the speed of sending messages. They knew that current flows very quickly along a wire, much more quickly than even the fastest operator can use a machine to send messages by hand. One way of getting higher speeds was to use the fact that current could flow one way round the circuit, or the other. A flow in one direction could be the same as a dash in Morse, and a flow in the other could be the same as a dot. This system is called 'double current working'. It is faster than normal, or 'single', current working. A dash or dot is signalled for the same length of time, but in ordinary Morse code a dash is signalled by an impulse lasting three times as long as a dot.

A machine was invented to receive messages by making a mark on **a paper tape** with dots on one side of a line and dashes on the other. (See figure 3.) Through the years other machines

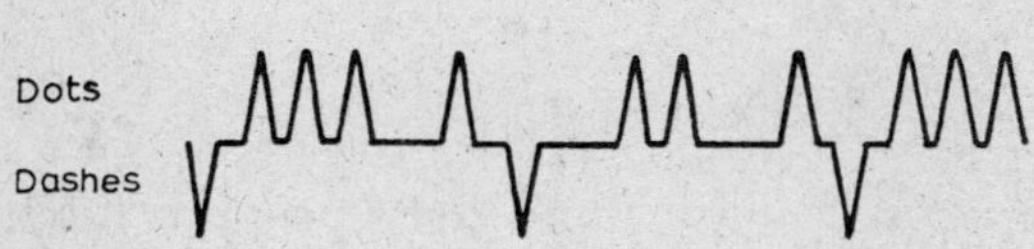

Figure 3

were invented, both to receive and to transmit messages. These machines are called **teleprinters.** They use a code called the International 5-unit teleprinter code instead of the Morse code. It is called a '5-unit' code because every letter or figure is made up of five impulses of electrical current, either positive or negative – three positive, two negative; one positive, four negative; and so on. All the letters are the same length in time. This is quite different from the Morse code, where E is the shortest and 'nought' the longest.

A teleprinter is really a machine which sends a typewriter message over a telegraph circuit. In fact, it looks like a typewriter and has the same keyboard. Every time one of the keys on the keyboard is pressed it sends the five impulses that make one letter or figure. It also sends a 'start' impulse. This sets the teleprinter (or any number of them that are connected) at a **receiving station** ready to print the letter that follows immediately. A 'stop' impulse brings all the teleprinters to rest after each letter or figure.

In the early days of **telegraphy** direct current was used. This is current which flows along a line in one direction at a time. Most modern telegraph systems use alternating current. This is current which flows backwards and forwards along the wire. To show that a current is alternating we draw it as in figure 4.

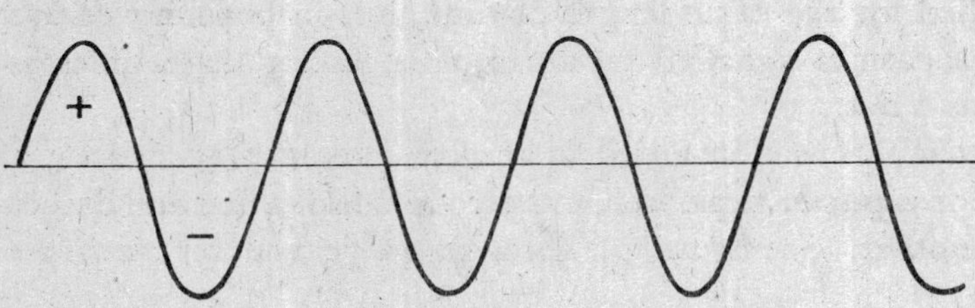

Figure 4

Alternating current is usually supplied to people's houses at 50 **cycles** per second. This means that every second the current **reverses** its direction 100 times. Alternating current used to send telegraph messages reverses its direction much more **frequently** than this. We say that it works on much higher frequencies. These frequencies range from 420 cycles per second to 3,180 cycles per second.

As the number of people who wanted to send telegraph messages increased, more and more messages **had to travel** along the same telegraph line. Using alternating current, one line can be used to send many messages at the same time if each message is sent out at a different frequency. A **transmitting machine** sends a message at one of the high frequencies mentioned, say 420 cycles. A receiving machine also working on a current alternating at 420 cycles per second will get the message. But it will not get a message being transmitted at the next highest frequency (which is 540 cycles per second) by another machine, even though this message is travelling along the same line at the same time. To get this message another receiving machine, working on current alternating at 540 cycles, must be used.

Adapted from *Over to You* by Roy Herbert
(Brockhampton Press)

Explanation of words in bold type

a circuit	is something that something else goes around, or **circulates** around.
interrupt	to stop for a short time.
the buzzer	is an instrument that makes a noise like the noise 'buzz' made by a bee.
increase	make greater.
a paper tape	a thin strip of paper.
teleprinter	a machine for printing messages which travel far ('tele-' – far). tele/graph; tele/printer; tele/vision; tele/gram; tele/phone.
receiving station	a place where messages are received. (The place from which the messages are sent is called a **transmitting station.**)
telegraphy	is the name for the system of sending messages by telegraph wire.
cycles	a cycle is a completed operation, in this case the flow of the current backwards and forwards once.
reverses	goes back on.
frequently	often.
had to travel	past for 'must travel'. I must travel to London today. OR I have to travel to London today. (present tense) OR I had to travel to London yesterday. (past tense)
a transmitting machine	sends, or **transmits,** a message. (A receiving machine, or a 'receiver', receives a message.)

Exercises

A Use these words to complete the following sentences: circuit, direct, direction, frequencies, increase, length, received, reverses, transmitted, transmitting.

1 Electrical impulses will not flow through a wire if the .... is broken.

2 Electricity can flow in either .... around a circuit.

3 New codes were invented to .... the speed of sending telegraph messages.

4 In the International 5-unit teleprinter code all the letters take the same .... of time to transmit.

5 In the Morse code, E can be .... more quickly than 'nought'.

6 Current which travels along a line in one direction only is called .... current.

7 Alternating current supplied to people's houses usually .... its direction 100 times every second.

8 Current used in telegraphy alternates at much higher .... than current supplied to people's houses.

9 A message sent by a .... machine working on current alternating at 540 cycles per second can only be .... by a machine working on the same frequency.

B Answer these questions.

1 How is a key used to send an electrical impulse through a circuit?

2 What is an 'earth return'?

3 How can the circuit shown in figure 2 be closed?

4 Why could Morse code be transmitted more quickly by 'double current working' than by 'single current working'?

5 Which sign takes the longest to transmit by Morse code?

6 How is a teleprinter being used to receive a message started?

7 What kind of current is mostly used in telegraphy nowadays?

8 What is the highest frequency used to transmit messages by alternating current?

9 Why was it necessary to increase the number of messages that could be sent along one telegraph line?

10 Why are machines now used instead of human operators to transmit and receive telegraph messages?

C Use each of these words in a sentence:

1. current 2. impulse 3. invented 4. connected 5. increased

3. The Telephone

We use the electric telegraph to send written messages to people far away from us. We use the telephone to talk to people far away. In many ways the telephone is better than the telegraph as **a means of communication.** The cost of sending a telegraph **depends on** the number of words in it. We have to make our telegraph message as short as possible, but, even in a short **telephone call,** we can say a lot of words. **A telegram** can only be sent from one post office to another. There is a delay before it can reach the person it is addressed to. The telephone **connects** you to a person directly. You may have to wait several hours for an answer to a telegram. You can ask a person questions and get the answers immediately **on the telephone.**

Sound travels through the air in waves. When you play a violin, for example, the violin string **vibrates.** The **vibrations** from the violin string pass through the air in little waves. When these waves reach the ear, **the eardrum** vibrates, and so you hear the violin. Different notes have different distances between the tops of the waves. We call these different 'wave-lengths'.

In 1875 an inventor called Alexander Graham Bell found that a thin sheet of metal, called a 'diaphragm', would vibrate in the same way as the eardrum when sounds reached it. He got a U-shaped iron magnet and wound coils of wire around it.

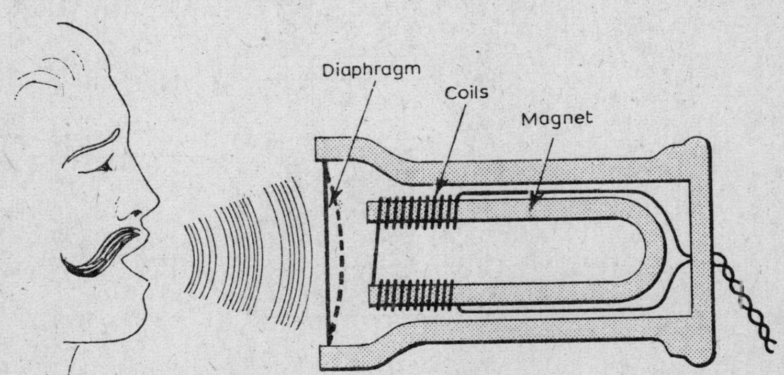

Figure 5

Then he placed the diaphragm very close to the **poles** of the magnet. (See figure 5.) Bell made sound waves reach the diaphragm, which vibrated, moving inwards towards the magnet and outwards from it. This made small currents of electricity pass through the coils and these currents were sent along a wire.

At the other end of the wire Bell placed a similar instrument, with a diaphragm and coils round a U-shaped piece of iron, which we call the 'receiver'. The impulses of electric current flowed through the coils of the receiver and **magnetised** the U-shaped piece of iron. The strength of the magnet was large or small according to the strength of the current. It made the diaphragm vibrate and the vibrations made waves of sound in the air exactly like the sound waves which **originally** reached the instrument at the other end of the wire. The sound waves had been turned into electricity, transmitted along a wire, and turned into sound again. The telephone had been invented.

But the sound from Alexander Bell's telephone could only be transmitted over short distances because the microphone was not very strong. A modern telephone has a carbon microphone in the part we call the **mouthpiece.** The diaphragm is still there, and when you speak into it, the waves of sound push it in

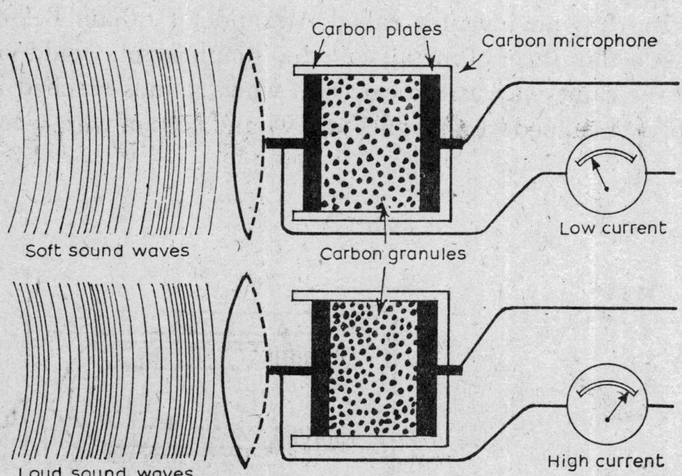

Figure 6

and out. But there is a current of electricity, supplied by a battery, which is already flowing through the microphone.

Behind the diaphragm are small pieces, or granules, of carbon. (See figure 6.) When you speak, you make louder and softer sounds. The louder or softer the noise you make, the more or less the diaphragm is pushed in or out.

Pushing in the diaphragm packs the carbon granules closer together. A soft sound does not push the diaphragm forward so far and the carbon granules are allowed to space out. Then it is more difficult for the current to flow through them and not so much gets through to the telephone line. The waves of electric current, varying like this, pass along the line and finally reach the receiver of the telephone held by the person you are speaking to. This receiver has an electro-magnet and a diaphragm and works just like the one Alexander Bell invented.

The line from your telephone is connected to the line of the telephone of the person you want to speak to through **the telephone exchange.** In the early days of the telephone, operators working in the exchange made all the **connections** between **callers** by hand. Nowadays, more and more exchanges are operated automatically.

Adapted from *Over to You* by Roy Herbert
(Brockhampton Press)

Explanation of words in bold type

a means of communication a way of making contact with somebody, by letter, for example, or on the telephone.

Telecommunications – making contact with someone far away. The telephone and the telegraph are different kinds of telecommunications.

depends on Are you going swimming today? It **depends on** the weather. (If it is warm I shall go. If it is cold I shall not go.)

telephone call When we use the telephone to contact someone we make a **telephone call.** Also, as a verb, 'Call him on the telephone.'

a telegram a telegraph message.

connects joins. Noun – connection. 'I can't hear you very well on this telephone. This is a bad connection.'

on the telephone We also say, 'I heard him **on** the radio.' 'I saw him **on** television.'

vibrates moves very quickly backwards and forwards. The wings of a fly vibrate when it is flying. Noun – **vibrations,** the quick movements backwards and forwards.

the eardrum is inside the middle of the ear. It is the most important part of the ear for hearing.

poles the two ends of the magnet where the magnetic power is concentrated.

magnetised turned the iron into a magnet, gave it magnetic power.

originally at first. Opposite – **finally** – in the end.

the mouthpiece the part of the telephone that we speak into.

the telephone exchange the building where telephone connections are made between the telephones of people who want to speak to each other.

connections see 'connects' above.

callers those who are telephoning.

Exercises

A Use these words to complete the following sentences: call, carbon, communication, connections, electro-magnet, magnetise, mouthpiece, on, transmit, vibrated.

1 The telegraph and the telephone are means of . . . .

2 When we want to speak to someone far away we make a telephone . . . .

3 He called his friend . . . . the telephone.

4 The diaphragm in Alexander Bell's invention . . . . when sound waves reached it.

5 Electric current through a coil of wire around a piece of iron will . . . . the iron.

6 The microphone in Bell's telephone was not strong enough to . . . . sound very far.

7 We speak into the . . . . of the telephone.

8 Electricity flows more easily when the . . . . granules are pushed together.
9 The receiver of a telephone contains an . . . . and a diaphragm.
10 Nowadays, ... between callers are usually made automatically at telephone exchanges.

B Answer these questions.
1 Give two reasons why the telephone is a better means of communication than the telegraph.
2 Why do we make telegraph messages as short as possible?
3 How does sound pass through the air?
4 What are the differences between the tops of sound waves called?
5 What happened when the diaphragm in Alexander Bell's telephone vibrated?
6 How was the U-shaped piece of iron in Bell's receiver magnetised?
7 What does the diaphragm do to the carbon granules in a modern telephone?
8 When does electricity pass more easily through the carbon granules?
9 What happens to the sound waves of your voice when you speak into the microphone?
10 Where is your telephone line connected to the line of the person you want to speak to?

C Use each of the following words in a sentence:
1. receiver 2. transmit 3. flows 4. vibrate
5. concentrated

4. Radio

In Communications 1 you read about electricity leaking away from underwater cables. In the early days of the telegraph and the telephone, men **observed** that signals sometimes jumped across the space between wires running next to each other, so that messages sent to one person went to another. They **realised** that electricity could pass through the air as well as along a wire.

Just as sound travels through the air in waves, so electricity travels through the air in what we call radio waves. Marconi, an Italian inventor, had the idea of sending messages by radio waves, or **'by wireless'.** He made a transmitter with a key to open and close an electrical circuit. His receiver was a tube containing metal filings with a battery connected to one end of it and a telegraph buzzer connected to the other end. The metal filings allowed an electrical current from the battery to pass through them when radio waves arrived from the transmitter. This current operated the buzzer and so Marconi was able to send Morse code messages like this.

Gradually Marconi **improved** his equipment until he could send messages greater and greater distances. In 1901 he wanted to send a wireless message across the Atlantic ocean from Cornwall, England, to Newfoundland. At that time people believed, quite rightly, that radio waves travelled in straight lines. As they knew that the surface of the earth is curved, they thought that Marconi's message would go right out into space and not reach Newfoundland. However, on 12th December, 1901, Marconi, in Newfoundland, heard three short dots (the Morse code signal for 'S') transmitted from his station in Cornwall, 1,800 miles away.

In fact, the radio waves had travelled in straight lines, but they had been **reflected** from the sky back to the earth's surface. This is how radio waves travel around the world. You have heard **on the radio** of 'long-wave **transmissions**' and 'short-wave transmissions'. In fact, there are two ceilings in the sky, not solid ceilings like the ceiling in a room, but more like nets which allow some radio waves to pass through them and reflect others. The lower ceiling is called the 'Heaviside layer'. This reflects long waves. Short waves are reflected by the higher ceiling, called the 'Appleton layer'. Radio **transmitters** can also send out **ultra-short waves.** These pass through both layers in the sky but are useful for **communicating** with space ships and can even be used to send sound very long distances when they are reflected from satellites in space.

It was exciting to be able to communicate over long distances by signals sent by wireless, but soon men wanted to try to

broadcast the sound of voices and even music in this way. A modern broadcast transmitting station has an aerial from which sound waves combined with **carrier waves** are transmitted. These waves are received on the radio sets in people's houses.

However, broadcasting like this would not be possible without the use of valves in the transmitting station and inside the radio sets. Radio waves are sent out by alternating current. Valves at the transmitting station are used to keep the electric circuits oscillating, that is, sending the current backwards and forwards in the aerial. The number of times the current alternates decides the **frequency,** or **wave-length,** on which the station is broadcasting.

Valves inside radio sets oscillate the current from the radio waves which are received. By turning a switch you can arrange that the oscillations of the circuits in your radio set are the same as those at one of the **broadcasting stations.** We say that you are 'tuning in' to that broadcast.

Valves inside your radio set have two other uses. Firstly, they change alternating current into direct current, and it is direct current that goes to the **loudspeaker** in the radio. It makes the centre of the loudspeaker vibrate to produce the sound waves that you hear. Secondly, the valves **amplify** the very small amount of electrical energy received from the transmitter. This energy would be far too little to be any use if it were not amplified by a valve.

Adapted from *Over to You* by Roy Herbert
(Brockhampton Press)

Explanation of words in bold type

observed	saw by watching carefully. Observer – a man who watches carefully.
realised	came to understand.
wireless	it has no wire (cf. use**less** – it has no use).
improved	made better.
reflected	sent back.
on the radio	cf. 'on the telephone'.

transmissions to transmit is to send out. A **transmitter** is something that sends out messages. Transmissions are the messages that are sent out.

ultra-short waves are very much shorter than short waves.

communicating sending messages to and receiving messages from.

broadcast programmes sent out over a wide area are **broadcast** from **broadcasting stations.**

carrier waves help to carry the sound waves through the air.

frequency refers to how often the electric current is sent backwards and forwards, how many times per second this is done. You may hear a radio announcer say, 'This station is broadcasting on a frequency of 800 kilocycles per second.'

wave-length is the distance between the tops of the sound waves that are broadcast. An announcer may say, for example, 'This station is broadcasting on a wavelength of 400 metres.'

loudspeaker this is used to change electric waves into sound waves and to make radio messages and radio programmes sound louder so that many people can hear them at the same time.

amplify make stronger.

Exercises

A Use these words to complete the following sentences: air, alternating, amplify, buzzer, Heaviside, higher, space, transmit, tune, wireless.

1 Electricity passes along a wire, but it may also pass through the . . . .

2 Marconi was one of the first people to send messages by . . . .

3 The . . . . in the receiver that Marconi invented sounded when an electric current reached it.

4 People thought that Marconi's message to Newfoundland would disappear into . . . .

5 The Appleton layer is the name for the . . . . ceiling in the sky which reflects radio waves.

6 Long waves are reflected by the . . . . layer.
7 Ultra-short waves are used to . . . . messages to space ships.
8 You can . . . . in to a broadcasting station by turning a switch on your radio.
9 . . . . current is used to transmit radio waves.
10 Valves inside your radio . . . . the electrical energy received from the transmitter.

B Answer these questions.

1 In the early days of telegraphy, why were messages sometimes sent to the wrong person?
2 What did Marconi use to make his first telegraph receiver?
3 Why did people think that Marconi's telegraph message would not reach Newfoundland?
4 What happens to radio waves when they go into the sky?
5 What is the Heaviside layer?
6 How can we send messages to space ships?
7 How do satellites in space help us to broadcast?
8 From what part of a broadcasting station are sound waves transmitted?
9 What are valves used for at a broadcasting station?
10 Give *two* things that the valves inside a radio set do.

C Use each of the following words in a sentence:

1. magnetise 2. invented 3. automatically 4. teleprinter 5. microphone

Metals

1. Metals and Civilisation

Long before men used metals they knew how to grow food, rear animals, build houses, make **pots,** and shape stones. Metal working was not the first craft known to man, but our knowledge of metals, and our use of metals, has been very important in forming our present material civilisation.

The first users of metals knew that they were bright materials, which were hard, but could easily be hammered into useful shapes. All the metals seemed very heavy. Iron, for example, was nearly three times as heavy as **granite.** If the metal worker was also a hunter or a soldier he was pleased with the sharpness of metal weapons and with their strength to last for a long time. Craftsmen who made things of wood or stone found that metal **cutting-tools** made their job easier.

Men learned to make fires and, later, to build furnaces hot enough to **liquefy** the metals they used. They found that the **molten metal** could be poured into **cavities** of stone or clay and left to become solid. Then the metal became the shape of the **mould** cavity. This liquefying process is called casting.

Men learned to **blend** metals to form an **alloy.** This alloy was sometimes stronger and harder than the metals that were used to make it. Probably the first alloy that was made was bronze, made of copper blended with about one part in ten of tin. If men used a **higher content** of tin the alloy was harder. If they used less tin the alloy was softer. So men made bronze with different contents of tin **depending upon** what they were making. The Romans who came to Britain used iron and bronze for weapons, tools, and farming implements; copper for vessels and ornaments; lead for water pipes, baths, and coffins; tin, gold, and silver for ornaments; and silver, brass, and bronze for coins.

Men called gold and silver 'noble metals' because they did not **tarnish** when **exposed** to the atmosphere for a long time, and because they could be melted again without much loss of weight. So these metals were used for making jewellery and for making coins. They became a measure of wealth. Men kept gold and silver to become wealthy. They did not use these 'noble metals' for making other things. They called all the other metals they used for making things 'base metals'. These 'base metals' included tin, lead, copper, and iron.

All metals are similar in some ways. Some men thought that they could change one metal into another. They wanted to change base metals into noble metals, in order to become wealthy. They tried many experiments to make this change. These men were called 'alchemists'. They called their experiments 'the science of alchemy'. The experiments were always unsuccessful.

But metal workers were making successful experiments. They learned how to change earthy minerals, called **'ores'**, into metals, by **smelting** them with charcoal in a fire or **furnace.** They found out which ores could be smelted and how many operations were needed to change them into metal.

Later, men learned how to make bigger and more efficient furnaces to make greater quantities of metal. Two hundred years ago the total production of iron in Britain was only 17,000 tons a year. A hundred and twenty years ago it was 600,000 tons. In 1952 the total production was over ten million tons.

In 1856 a man called Henry Bessemer told people about a method he had discovered to change large quantities of **crude pig iron** into steel. Soon afterwards, a man called Charles Siemens found another good method of making steel. Steel was especially useful for making bridges, railway lines, ships, guns, and, later, motor cars.

The art of metal working developed into the science called **metallurgy.** In 1861 Professor Sorby began examining metals through microscopes and started that branch of metallurgy called **metallography.** He studied the properties of metals, their melting points, their strength, and their hardness. A new light metal called aluminium was discovered. Some nickel was blended with steel to form alloy steel, better than ordinary steel for many purposes.

Today many of the things we use are made of metal. The metallurgist must make just the right kind of metal for each thing. An alarm clock with metal working parts wakes us in the morning. We press an electric switch and current passes along a copper wire to light a lamp with a **tungsten filament.** We wash in water that has come through copper or lead pipes. We shave with a hardened-steel razor blade. Tea is made in a metal tea-pot. Water is heated in an aluminium kettle. We eat our breakfast with a **stainless-steel** knife and fork. We read a newspaper printed from a lead alloy **type-metal.** We go to work on a bus made of metal. We pay our fare for the bus with copper–nickel alloy or copper–tin–zinc alloy coins. Many of the things we use at work, for example, lathes, typewriters, or tractors, are made of metal.

In the past men lived in ages called after a single new material – the Stone Age, the Bronze Age, or the Iron Age. Today we use many different kinds of materials. The metallurgist is a scientist who decides which material is most suitable for each purpose. Our age may be called the 'Early Scientific Age'.

Adapted from *Metals in the Service of Man* by William Alexander and Arthur Street (Penguin Books Ltd.)

Explanation of words in bold type

pots	bowls, jugs, vases, etc. (potter – a man who uses clay to make bowls, jugs, and vases.)
granite	a heavy kind of stone.
cutting-tools	knives, axes, etc.
to liquefy	to change something into a liquid.
molten metal	metal which has melted. (He **melted** the metal. It became **molten** metal.)
cavities	hollow spaces. (A **cave** is a hollow space in the side of a hill.)
a mould	is used to form something into a shape. (verb – to mould. The potter **moulded** the clay into the shape of a bowl.)

to blend metals	to mix different metals together.
alloys	two or more metals blended together (e.g. steel = iron + carbon, brass = copper + zinc).
alloy steel	steel blended with an element, such as nickel, or tungsten.
higher content	What are the **contents** of bronze? (What does bronze contain? What is bronze composed of?) The **contents** of bronze are copper and tin. (It contains copper and tin. It is composed of copper and tin.) The copper **content** in bronze is **higher** than the tin content. (It contains more copper than tin. It is composed of more copper than tin.)
depending upon	Will this knife cut wood? It **depends upon** how hard the wood is.
to tarnish	to form oxides on the surface.
exposed	left open to. (Milk exposed to hot weather soon goes bad. Iron exposed to a damp atmosphere will tarnish.)
ores	ores contain minerals from which metals can be taken. Ores almost always contain earthy matter as well as minerals. Iron ore – the form in which iron is found in the ground.
smelting	changing ore into metal by heating.
furnace	a chamber in which ore is heated by a fire.
crude pig iron	iron that is not pure. It is made in the furnace but still contains carbon and other impurities.
metallurgy	the science of making metals, of blending metals to form alloys, of treating metals with heat.
metallography	the study of the structure of metals, usually with the aid of a microscope.
tungsten filament	a thin thread of a metal (tungsten) with a high melting point. It is made hot in an electric lamp to give light.
stainless steel	steel that will not tarnish when exposed to the atmosphere.
type-metal	a special metal suitable for printing letters clearly.

Exercises

A Use these words to complete the following sentences:
alloy, copper, current, furnace, higher content, metallography, metallurgist, mould, smelted, steel, tarnish, tin, tungsten.

1 Metals are melted in a . . . .
2 Molten metals are formed into shapes in a . . . .
3 Metals are blended together to form an . . . .
4 Bronze contains . . . . and . . . .
5 There is a . . . . of copper than of tin in bronze.
6 Iron will . . . . if left exposed to a damp atmosphere.
7 Ores are . . . . in a furnace to make metals.
8 When motor cars were invented they were made of . . . .
9 When we press an electric switch . . . . passes along a copper wire.
10 An electric lamp has a filament made of . . . .
11 A man who studies the science of metals is called a . . . .
12 The study of the structure of metals is called . . . .

B Answer these questions.

1 What things could men do before they learned how to use metals?
2 Why were hunters and soldiers pleased with metal weapons?
3 Why were craftsmen pleased with metal tools?
4 How did the first users of metals form them into shapes?
5 How did men liquefy metals?
6 What did they do with the molten metal?
7 How is an alloy formed?
8 How did men make bronze?
9 How did men make the bronze harder?
10 What metals did the Romans who came to Britain use?
11 Why were gold and silver called 'noble metals'?
12 Which metals were called base metals?
13 Why did some men want to change 'base metals' into 'noble metals'?
14 What did men use to smelt ores?
15 How much iron was made in Britain in 1952?

16 Give the names of two men who played an important part in the discovery of making steel.
17 What were the properties of metals studied by Professor Sorby?
18 Give the names of three metals used in our homes today, and say what they are used for.
19 Give the names of three things made of metal used by people at work.
20 Why might our present age be called the 'Early Scientific Age'?

2. Making Iron

Many years ago men first heated iron ore in a fire. When the fire went out a piece of iron like a **sponge** was left. This **spongy** iron could be hammered into shape to make tools and weapons. Men discovered that when the wind was blowing the fire burnt faster and hotter. Iron was produced faster. They made **bellows** to blow air on to the fire. The bellows blew a **blast** of air on to the fire.

Six hundred years ago men learned how to make a very strong blast of air for iron **production.** This made the fire in the furnace very hot. The iron became very hot and melted. The molten metal could then be poured as a liquid from the furnace. The furnace is called a blast furnace. The iron produced in the blast furnace is called 'cast iron' or 'pig iron'. It is more **brittle** than iron which has been heated but not melted, but it is easier to shape, as it can be poured into moulds. At first men used **charcoal** to heat the furnace. Nowadays **coke** is used.

A MODERN BLAST FURNACE

Look at figure 7. It is a **sectional diagram** of a blast furnace. A large furnace is about 100 ft. high and about 28 ft. in **diameter** at its widest part. It is covered on the outside with steel plates half an inch thick. On the inside is a lining of **firebricks.** Nearly a million firebricks are used for one blast furnace.

Iron ore, coke, and limestone are put into the top of the furnace. These are called the **'charge'.** The coke provides the fuel to melt the metal in the iron ore. When the metal is melted it falls to the bottom of the **hearth.** Limestone **combines with**

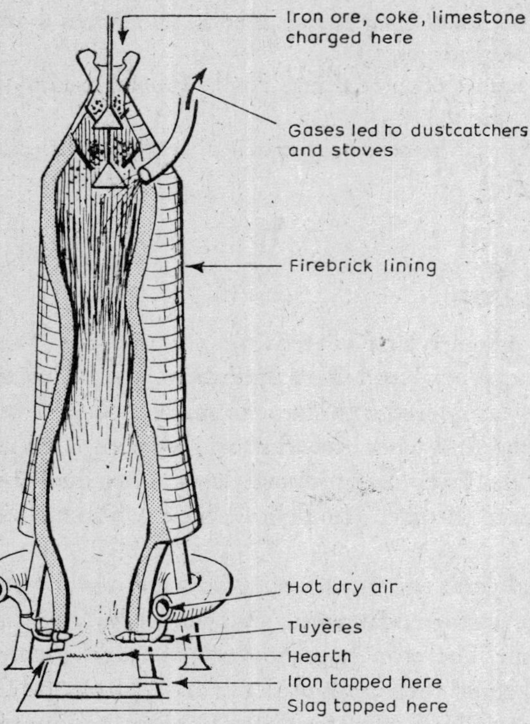

Figure 7 *A blast furnace*

the earthy matter in the iron ore. The limestone and the earthy matter combined form slag. When the temperature is very hot this slag liquefies. It falls to the bottom of the hearth. The slag is **less dense** than the molten metal. It forms a layer on top of the molten metal. There are two openings near the bottom of the furnace. When a certain quantity of molten metal and slag has formed in the hearth, the metal is **tapped** through the lower opening and the slag is tapped through the upper opening. The molten metal is **white-hot** when it runs out. Sometimes it runs into moulds to make bars of pig iron. Sometimes it runs to a **steel works** to be changed into steel.

A blast of hot air is blown into the bottom of the furnace to make the coke burn fiercely. It is forced into the furnace through **tuyères.** (See figure 7.) These tuyères are placed around the **circumference** of the blast furnace about 8 ft. above the bottom.

While the coke is burning and the iron is melting, gas is formed at the top of the chamber. This gas is led off from the top of the furnace to be used. It contains carbon monoxide, which is **combustible.** Part of this gas is used for making the air blast hot. It is led off into stoves. (See figure 8.)

These stoves are nearly as large as the blast furnace. There are usually three stoves with each blast furnace. (The third is a **spare** stove.) Look at figure 8. In the first stove the carbon

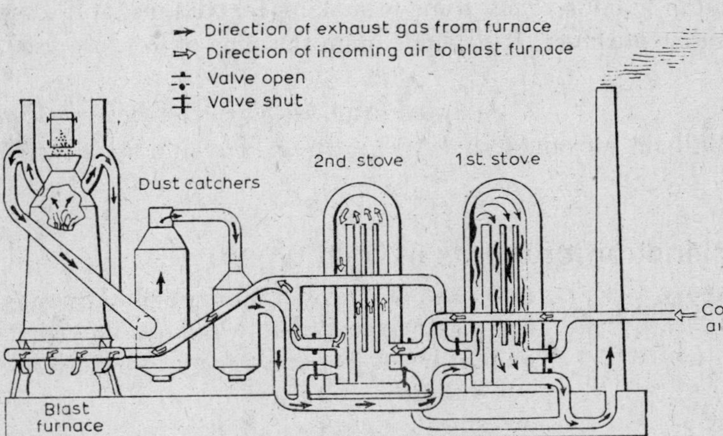

Figure 8 *The stoves of the blast furnace*

monoxide in the gas from the furnace is being burnt. This burning gas is led through rows of firebricks. It makes the firebricks very hot. The rows of firebricks in the second stove have already been made very hot like this. A blast of cold air is blown into this second stove. It becomes hot air as it passes through the stove and it is then blown into the furnace. After half an hour a change is made. Cold air is led into the first stove to be heated. Gas from the furnace is led into the second stove to be burnt and to **re-heat** the firebricks. Every half-hour a change is made like this. In figure 8 you will see that a dust catcher is used to take the dust out of the gas before it is led into a stove.

When a hot blast of air is used to make the coke burn fiercely, less coke is needed than when a cold blast of air is used. In some furnaces a cold blast of air is still used to produce a **high-grade** iron. This high-grade iron has less of certain impurities than iron produced with a hot blast.

The gas that is not used to make the air blast hot is not wasted. It is led away to be used for driving pumping engines or for providing electricity for light and power in the iron works, or even in a near-by town.

The slag from the blast furnace is not wasted. Some slag is useful in making roads, some in making **fertilisers,** and some for other purposes. It depends upon the type of iron ore used.

Adapted from *Metals in the Service of Man* by William Alexander and Arthur Street (Penguin Books Ltd.)

Explanation of words in bold type

a sponge	is soft and has many holes through it. **Spongy** iron – a soft piece of iron with holes through it.
bellows	an instrument that draws in air and blows it out with force.
blast	air blown with force.
production	making something. Iron production – making iron. The iron is produced in an iron works.
brittle	something that breaks easily is brittle.
charcoal	is black and has been made by burning wood slowly, not allowing much air to get to it.
coke	grey brittle matter, containing about 80% carbon. It is left when coal is heated and the coal-gas drawn off.
sectional diagram	a diagram to show how something would look if one side were cut off. In figure 7, you can see both the inside and the outside of the blast furnace.

diameter of the blast furnace the length of a line passing from one side to the other through the centre.

firebricks bricks that will not be damaged by heat in the furnace.

the charge the things that are put into the furnace to make the metal. The furnace is **charged** with iron, coke, and limestone.

hearth a place to contain fire.

combines with mixes with. Limestone combines with the earthy matter of the ore to form slag. **Slag** – the limestone and the earthy matter combined.

less dense it has less mass per unit volume. Cement is more dense than sugar.

tap to take the molten metal or liquid slag from the furnace in a controlled manner.

white-hot metal very hot metal that has turned white.

steel works a place for making steel. He works at the steel works.

tuyères (pronounced 'tweers') – small openings through which the blast is forced into the furnace.

circumference of blast furnace the distance around the blast furnace.

combustible it will burn. Combustion – burning.

spare something not being used, but ready to be used if necessary. The car has a spare tyre in the back in case one tyre gets a puncture.

re-heat to heat again. Your dinner has been cooked but it is now cold. I will re-heat it.

high-grade good quality. High-grade petrol costs more than low-grade petrol.

fertilisers are used to make land fertile. They are spread upon the earth or dug into the earth. When land is useful for growing things it is called **fertile** land.

Exercises

A Use these words to complete the following sentences: bellows, blast, bottom, brittle, coke, dust catcher, gas, firebricks, hearth, less, limestone, melt, steel plates, stove, tapped, top, tuyères, wasted, works.

1 Many years ago . . . . were used to blow air on to the fire to make iron.
2 When metals become very hot they . . . .
3 A strong force of air is called a . . . . of air.
4 Something which breaks easily is . . . .
5 The outside of a furnace is covered with . . . .
6 The charge is put into the . . . . of the furnace.
7 The fuel used in the furnace to melt the metal is . . . .
8 The earthy matter in the iron ore combines with . . . . to form slag.
9 The molten metal and the liquid slag fall to the bottom of the . . . .
10 The metal and the slag are . . . . through openings near the . . . . of the furnace.
11 Steel is produced in a steel . . . .
12 Hot air is forced into the furnace through . . . .
13 . . . . is led off from the top of the furnace.
14 Some of the gas is used to heat . . . . in the stove.
15 Cold air is blown through the . . . . to be made hot.
16 Dust is taken out of the gas in a . . . .
17 . . . . coke is needed when a hot blast of air is used than when a cold blast is used.
18 The gases and the slag from the furnace are used. They are not . . . .

B Answer these questions.

1 What happens when wind blows on to a fire?
2 Why is a furnace for making iron called a 'blast' furnace?
3 How is iron made into shapes after it has been heated?
4 How many firebricks are used for one blast furnace?
5 What materials are used to charge the furnace?
6 Why does the liquid slag form a layer on top of the metal?
7 *Two* things may happen to the molten iron from the furnace. What are they?
8 How far from the bottom of the furnaces are the tuyères?
9 What happens to the carbon monoxide in the stove?
10 Why is a hot blast of air used more often than a cold blast in furnaces?
11 What may happen to the gas that is not led into the stove?
12 Give two things that the slag from the furnace may be used for.

C Write sentences using each of the following words or phrases:

1. mould 2. alloy 3. iron ore 4. it depends upon
5. melted.

3. Making Steel

There are two main kinds of iron used for making steel – 'basic' iron and 'hermatite' iron. These kinds of iron can be called steel-making irons. Both contain small quantities of carbon, silicon, sulphur, phosphorus, and manganese. The **percentage** of these elements may be **approximately**:

	In basic iron	In hermatite iron
Carbon	3·5%	3·75%
Silicon	0·85%	2·0%
Sulphur	0·08%	0·04%
Phosphorus	1·6%	0·045%
Manganese	1·0%	0·5%

Basic iron contains more phosphorus but it contains less silicon than hermatite iron.

Steel made from basic iron is called 'basic steel', and steel made from hermatite iron is called 'acid steel'.

The **main difference** between iron and steel is in the carbon content. Iron contains from 3 to 4% carbon. In making steel the carbon content is **reduced** to between 0·07 and 1·4%.

There are several different **grades** of steels. We may **classify them** according to their carbon content.

Low-carbon steels:	Less than	0·07–0·15% carbon
Mild steels		0·15–0·25% carbon
Medium-carbon steels		0·25–0·5 % carbon
High-carbon steels		0·5 –1·4 % carbon

(There are also alloy steels which we shall learn about later.)

The low-carbon steels are called 'soft' and the high-carbon steels are called 'hard'. When making something you choose the best steel for the purpose. For example, soft steels are used for making wire, **rivets,** and sheet steel. Hard steels are used for making springs, hammers, and chisels. Mild steel is usually used for general engineering purposes.

To make steel from iron it is necessary to reduce the carbon content. Here are the approximate percentages for carbon, silicon, sulphur, phosphorus, and manganese in basic iron, and in mild steel:

	Basic iron	Mild steel
Carbon	3·5	0·15
Silicon	0·85	0·03
Sulphur	0·08	0·05
Phosphorus	1·6	0·05
Manganese	1·0	0·5

You will see that other elements as well as carbon have been reduced, especially the silicon and phosphorus. These elements are called **'impurities'** in the metal. How are they reduced? The process by which they are reduced is called 'oxidation'. The two main processes used to turn iron into steel are called: the Bessemer Process, and the Open Hearth Process.

THE BESSEMER PROCESS

Henry Bessemer first told people about the process named after him in 1856. In his process the impurities in the metal are burnt away by blowing air through molten iron. The oxygen in the air combines with some of the iron to produce iron oxide. The iron oxide 'oxidises' the silicon and the manganese and these, together with the iron oxide, form slag. The iron oxide also oxidises the carbon, which forms carbon monoxide and carbon dioxide gas. This is why the process is called 'oxidation'. The silicon, manganese, and carbon are oxidised. When these impurities have formed slag or gas they can be separated from the iron. The metal that is poured away is steel.

The Bessemer Process uses a furnace called a **'converter'.** (See figure 9.) The outside of the converter is made of steel plates. The inside is lined with bricks. The converter is tipped on to its side and the **charge** of molten iron is poured into the top. Then the converter is put upright again. A blast of air is blown through holes in the **base** of the converter. This is called the 'blow'. The air forces its way through the molten metal. The oxidation process begins. This oxidation forms heat. The

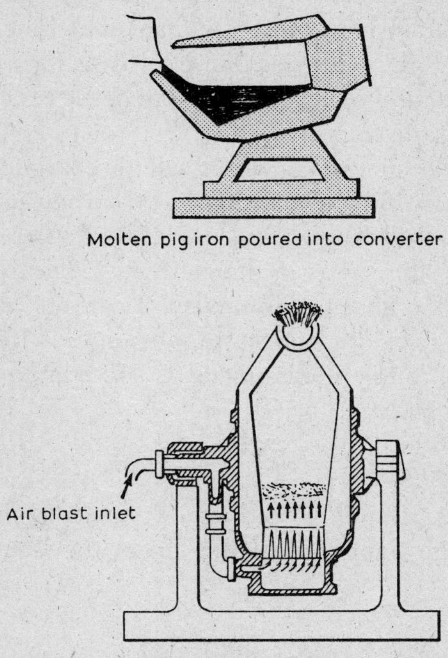

Figure 9 *The Bessemer converter*

temperature rises from 1,300° C. at the beginning to 1,600° C. at the end of the process. No outside heat is used.

When the blow begins a shower of sparks comes from the top of the converter. These change into a sheet of flame which goes 9 m. into the air and makes a roaring sound. The carbon is being burnt out of the top of the converter in the form of carbon monoxide and carbon dioxide gas. The manganese and silicon are being burnt and changed into slag inside the converter. The process takes about twenty minutes. At the end of that time the flame dies down and brown smoke comes out of the top of the converter. After another three or four minutes the slag is poured into a **ladle.**

(Henry Bessemer did not know how to take the phosphorus out of the iron. Phosphorus is an impurity in steel. In 1878 a charge of limestone was first put into the top of the converter as well as the charge of molten iron, and **dolomite** was put into the firebricks lining the inside of the converter. The limestone and the dolomite combine with the phosphorus to form slag. As you have already read, the slag is poured out of the converter before the metal is poured into the ladle.)

While the metal was in the converter, all the carbon was **removed** from the iron. But steel without any carbon in it at all would be too **brittle** to use. Also, oxides and gases have formed in the metal while it was in the converter. These must be removed. To add the necessary quantity of carbon, and to take out the unwanted oxides and gases, an alloy of **'ferro-manganese'** is added to the molten metal as it is poured from the converter into the ladle.

THE OPEN HEARTH PROCESS

In the Bessemer Process a lot of the heat used is wasted. It burns away and forms a high flame out of the top of the Bessemer

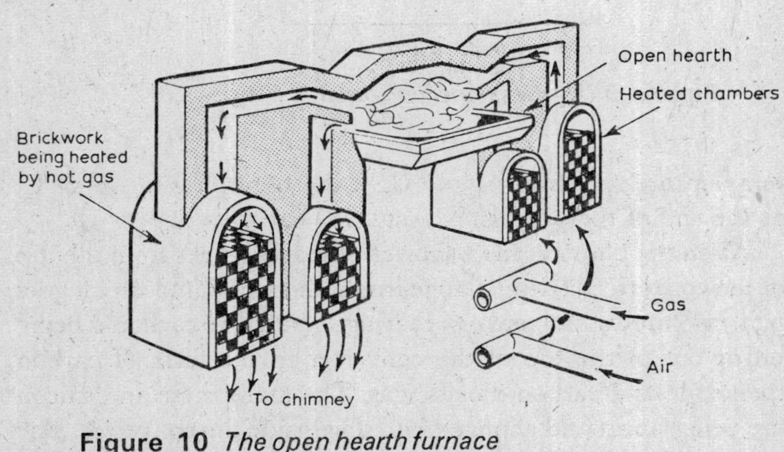

Figure 10 *The open hearth furnace*

converter. Soon after the Bessemer converter was first used, two brothers called Siemens found a way of saving the heat used in making steel. This heat was used again in the steel-making process. This process is called the Open Hearth Process.

A charge of iron and **scrap** steel is put into an open hearth. (See figure 10.) Gas and hot air are the fuel. They are led through openings into one side of the hearth. The gas and hot air combine above the hearth and burn fiercely. A hot gas is formed. This hot gas is led away out of the hearth through openings on the other side. It is then led through another **pair** of chambers which are filled with **chequered brickworks.** The hot gas heats the bricks. When the bricks are very hot the direction of the flow of gas and air is **reversed.** The gas and hot air are led into the hearth through the brickwork chambers which have been heated, and the hot gas is led out of the hearth to heat the other pair of brickwork chambers.

So when the gas and air come into the hearth they are already very hot. The hot gas is not wasted. It is used again. It heats the brickwork chambers that the gas and air will be led through afterwards.

Adapted from *Metals in the Service of Man* by William Alexander and Arthur Street (Penguin Books Ltd.)

Explanation of words in bold type

percentage the quantity out of a hundred. 10% – 10 out of a hundred.

approximately not exactly. There are **approximately** 300 students in the College. There are **exactly** 289 students in the College.

the main difference the most important difference.

reduced made less. Fifty students were ill. The number of students in the College was reduced to 239.

grades a measure of quality. High-grade petrol costs more than low-grade petrol.

classify them put them into the classes (groups) they belong to.

rivets are used for joining pieces of metal together. (Nails and screws are used for joining pieces of wood together.)

impurities in the metal elements in the metal not the same as the metal itself. They are not pure.
converter to convert is to change. A converter changes something into something else.
the charge the material put into a furnace.
the base the bottom.
dolomite a white solid found in large quantities, sometimes forming whole mountain ranges. (The Dolomite Mountains are in Italy.)
ladle There are different kinds of ladles. A cook uses a ladle to take soup out of a saucepan. The cook's ladle looks like a large spoon. The ladle in a steel works is very much larger.
removed taken out.
brittle something that breaks easily is brittle.
ferro-manganese an alloy of manganese and iron.
scrap steel steel that has been used to make something that is no longer wanted. When a car is no longer useful it may be broken up to provide scrap steel and other scrap metal. This scrap metal is used again to make new metal.
a pair two together (e.g. a pair of shoes).
chequered brickworks an arrangement of bricks with holes between them. (See figure 10.)
reversed turned in the opposite direction. He drove past the house. Then he reversed his car back to it.

Exercises

A Use these words to complete the following sentences: base, blow, brittle, chequered, converter, dolomite, hard, heat, ladle, low, molten, oxidation, reduced, reversed, scrap, steel, temperature.

1 The carbon content in iron is greater than the carbon content in . . . .

2 The carbon content is . . . . when steel is made from iron.

3 . . . . carbon steels are called soft steels. High-carbon steels are called . . . . steels.

4 When steel is made impurities in the iron are reduced by a process called . . . .
5 In the Bessemer Process air is blown through . . . . iron.
6 The furnace in the Bessemer Process is called a . . . .
7 The blast of air comes into the converter through holes in its . . . .
8 When the air is forced through the molten metal it is called the . . . .
9 In the Bessemer converter the . . . . rises to 1,600° C.
10 Limestone and . . . . combine with the phosphorus in the converter to form slag.
11 The metal from the converter is poured into a . . . .
12 Steel that contains no carbon at all is too . . . . to use.
13 The charge for the Open Hearth Process is iron and . . . . steel.
14 The hot gas coming out of the open hearth is used to . . . . the bricks in the chambers with . . . . brickworks.
15 When one pair of chambers is hot enough to heat the gas and air, the direction of the gas and air is . . . .

B Answer these questions:

1 What are the two main kinds of iron used in steel-making called?
2 What is the carbon content of basic iron approximately?
3 What is the main difference between iron and steel?
4 Give the names of three things made from hard steels.
5 What is the approximate percentage of silicon in mild steel?
6 How are the impurities in the metal burnt away in the Bessemer Process?
7 Name two elements that help to form slag in the Bessemer Process.
8 What is the outside of the Bessemer converter made of?
9 What process forms heat in the Bessemer converter?
10 What gases are formed by the carbon in the Bessemer Process?
11 What happens to these gases?
12 What happens to the phosphorus in the converter?
13 Why must some carbon be put back into the metal before it makes useful steel?
14 What fuel is used in the Open Hearth Process?

15 What happens to the hot gas when it comes out of the hearth in the Open Hearth Process?

C Write sentences using each of the following words or phrases:

1. produced 2. hearth 3. molten 4. carbon content 5. firebricks

4. Alloys

Different metals are blended together to make alloys. Usually there is more of one metal than of the other in an alloy. Bronze is an alloy that was first made thousands of years ago. It was made by blending copper with about **one part in ten** of tin. The metal workers heated ores containing copper with ores containing tin. The two metals in the ores melted and blended together to make bronze.

Nowadays, there are not many alloys that are made in this way. Usually the different metals are **produced** separately and all impurities are removed from them. Then the pure metals are sent to a factory where they are blended to make an alloy.

Most alloys are made by **mixing** together metals which have been melted. We say they are **molten metals.** When they are mixed together the **mixture** is put into a **mould.** Then it grows cool, and becomes solid.

Some alloys are difficult to make because the metals in them have different melting points. For example, brass is an alloy of copper and zinc. Copper melts at a temperature of 1,083° C. Zinc melts at 419° C. and it boils at 907° C. We cannot make brass simply by heating pieces of copper and zinc together to a temperature of over 1,083° C. At that high temperature both metals would certainly melt, but the zinc would boil away until nothing of it was left.

So how do men manufacture brass? First they heat the metal that has the higher melting point. This is the copper. When this becomes molten, the solid zinc is added to it. The zinc quickly **dissolves** in the molten copper before much zinc has boiled away. A little of the zinc boils away and is lost. Zinc must be about one part in twenty of the alloy. So a little **extra** zinc is put in with the molten copper to allow for the little that boils away.

In brass, the metal with the higher melting point (copper) forms **a greater part of** the alloy than the metal with the lower melting point (zinc). In some alloys the metal with the higher melting point forms **a smaller part** of the alloy than the metal with the lower melting point. For example, there is **a light alloy** which contains 92% aluminium (melting point 660° C.) and 8% copper (melting point 1,083° C.) It is not possible to make this light alloy by melting the small part of copper and then adding the much larger part of aluminium. So the **alloying** is done **in two stages.** Firstly, something called a 'hardener alloy' is made. This hardener alloy contains 50% aluminium and 50% copper, and it has a low melting point. Its melting point is even lower than that of aluminium. Secondly, the aluminium is melted and the correct amount of the hardener alloy is added to the molten aluminium. To make 100 kilograms of this light alloy (aluminium–copper alloy) we need 84 kilograms of aluminium to be melted first, and 16 kilograms of the hardener alloy added to it.

Alloys are made for several reasons, often to make something stronger than each of the metals that **combine** to form the alloy. To make a strong alloy it is not always necessary to use two strong metals. For example, copper and aluminium are both **fairly weak** metals. But if the correct proportions of copper and aluminium are blended together a strong alloy is produced. If 5% of the aluminium is added to 95% of copper, the alloy produced is twice as strong as copper. If 10% of aluminium is added to 90% of copper, the alloy produced is three times as strong as copper. But if more than 10% of aluminium is used the alloy begins to get weaker again. If 16% of aluminium is used the alloy produced is useless, because it is very brittle. Alloys containing copper and aluminium with a greater proportion of copper are often called 'aluminium bronze' (though the word 'bronze' was first used in **ancient times** to mean an alloy of copper and tin).

You have also read about an alloy containing copper and

aluminium where the aluminium forms the greater proportion. This alloy contains **up to 12%** of copper. It is not so strong as aluminium bronze, but it is much stronger than aluminium alone. Also, this alloy is almost as light as aluminium alone, so it is often very **useful in industry.**

So alloys may be produced for extra strength, or for extra lightness. They may also be produced to resist **tarnishing,** or to give **high electrical resistance,** or for several other reasons.

Adapted from *Metals in the Service of Man*
by William Alexander and Arthur Street (Penguin Books Ltd.)

Explanation of words in bold type

one part in ten	one tenth, or 10%.
produced	made. 'Produced' is the word usually used about things made in a factory. Aeroplanes are **produced** in an aircraft factory. Production – the making of something. When the war started there was an increase in aircraft **production.**
mixing	putting different things together to make one thing. The **mixture** – is made when the different things have been mixed together.
molten metals	metals that have been melted and are still in a liquid form.
a mould	is used to form something into a shape.
dissolves	melts and mixes with. Sugar dissolves in hot tea.
extra	something added.
a greater part	a larger proportion, more.
a smaller part	a smaller proportion, less.
a light alloy	an alloy that does not weigh much.
alloying	making an alloy.
in two stages	one part of the alloying is done before the second part.
combine	join together.
fairly weak	a bit weak, but not very weak.
in ancient times	thousands of years ago.

up to 12% not **more** than 12%, although it may be **less** than 12%.

useful in industry can be used in a factory or workshop or a building, where things are manufactured or built.

tarnishing forming oxides on the surface. When iron tarnishes, we say it **rusts,** or it has gone **rusty.** Rust is a red substance that often forms on the surface of the iron when it is exposed to rain, air, or sea water. The word **corroded** is also often used when the surface of any metal is harmed in this way. The surface is **corroded,** or, there is **corrosion** on the surface.

high electrical resistance something that does not allow an electric current to pass through it easily has a high electrical resistance.

Exercises

A Use these words to complete the following sentences: aluminium, aluminium bronze, boil away, brittle, copper, dissolve, hardener, higher melting point, low, molten, three times, zinc.

1 Most alloys are made by mixing together . . . . metals.
2 Brass is an alloy made of . . . . and . . . .
3 Copper has a . . . . than zinc.
4 If zinc was heated to 1,083°C. it would . . . .
5 Small quantities of zinc soon . . . . in molten copper.
6 A . . . . alloy is made in the first stage of alloying a greater part of aluminium with a smaller part of copper.
7 The hardener alloy contains 50% aluminium and 50% . . . .
8 This hardener alloy has a . . . . melting point.
9 Alloys of copper and aluminium with a greater proportion of copper are often called . . . .
10 An alloy made of 90% copper and 10% aluminium is . . . . as strong as copper.
11 If 16% of aluminium is used to make an alloy with copper, the alloy will be very . . . .

12 An alloy made of copper and aluminium with a greater proportion of aluminium is much stronger than .... alone.

B Answer these questions.

1 How did metal workers make bronze thousands of years ago?
2 Why are some alloys difficult to make?
3 What is brass made of?
4 When brass is made, what happens to the zinc when it is added to the molten copper?
5 Why is extra zinc put in with the molten copper?
6 What does the 'hardener alloy' you have read about contain?
7 Why is an alloy containing copper and aluminium with a greater proportion of aluminium often very useful in industry?
8 Give three reasons why alloys are produced.

C Use each of these words in a sentence:

1. blended 2. impurities 3. temperature 4. proportion 5. extra

5. Aluminium

A hundred years ago in France a new precious metal was being used to make the spoons and forks for the Emperor's table. This was aluminium. An ore which contains aluminium was first discovered at Les Baux, in France. We call this ore bauxite. Nowadays, bauxite is also **mined** in America, Africa, India, and Austria. Over 2 million tons of aluminium are now produced in the world every year, so it is no longer **considered** a precious metal. But it is a very useful metal, for several reasons:

1. Aluminium is a light metal, with a **density** about a third of that of steel or brass.
2. Although aluminium alone is not particularly strong, it can form a very strong light alloy when other metals are mixed with it.
3. Aluminium is a **good conductor of electricity.**
4. Aluminium is a **good conductor of heat.**

5. Aluminium has a **higher resistance to corrosion** than many other metals. A thin **film** of oxide forms on its surface and protects it.

USES OF ALUMINIUM

Because it is a light metal, aluminium is very useful in **aircraft production. Manufacturers** want to make aeroplanes as light as possible. For the same reason aluminium is used by manufacturers of buses, railways carriages, and boats.

Because aluminium is a good conductor of electricity it is often used for cables to carry telegraph and telephone messages. It is not such a good conductor of electricity as copper, but it is lighter. So it is used for **overhead** telegraph and telephone cables. At first, when it was stretched between telegraph poles it **sagged** too much. Then men put a **core** of steel wire through the centre of the cable to make it firmer.

Because it is a good conductor of heat, aluminium is used for all kinds of **cooking utensils,** such as saucepans and frying-pans. The food being cooked in these utensils soon becomes hot.

Aluminium is also used to make foil. Foil is metal that has been rolled into a very thin sheet. It is used for wrapping food, such as butter, to protect it, for making bottle tops, and for the 'silver paper' in cigarette packets.

DURALUMIN

There is an alloy made by adding small quantities of copper and magnesium to aluminium which is called duralumin. It was first made by a German metallurgist, Dr Alfred Wilm. He mixed the aluminium with the copper and magnesium. Then he wanted to make the alloy stronger. First he heated it. Then he cooled it quickly by putting it in water. He tested the alloy, but he was not pleased because it was not so strong as he had hoped that it would be.

A few days later he decided to test the alloy again. He was very surprised to find that it was now much stronger. Wilm made experiments to find out what happened to the alloy when it was kept for different lengths of time. He discovered that this alloy reached full strength in about four or five days. We call this rise in strength 'age hardening'. In 1909 Wilm gave the right to use his discovery to the Durener Metal Works at Duren. So the alloy is now called duralumin.

Duralumin is now used in making aeroplanes because it is both light and strong. Thin sheets of it cover the wings and the **fuselage** of aeroplanes. Unfortunately, duralumin has less resistance to corrosion than aluminium alone. It is important that the metal used for making aeroplanes should resist corrosion, especially for aeroplanes flying in warm, damp climates, and for **seaplanes,** where the metal comes into contact with salt water. To protect the duralumin, a thin layer of **pure aluminium** is added to each side of the duralumin sheet. This makes a **three-ply metal.** Men tested this three-ply metal to see if it had a high resistance to corrosion. They **exposed it to** salt water for a long time. They found out that even a very thin sheet of this three-ply metal was not damaged by corrosion after five years.

Adapted from *Metals in the Service of Man* by William Alexander and Arthur Street (Penguin Books Ltd.)

Explanation of words in bold type

mined dug out of the ground. A bauxite mine – a place under the ground from where men dig bauxite.

considered thought to be.

density the measurement of mass per unit volume.

good conductor of electricity a current of electricity passes through it easily.

good conductor of heat heat passes through it easily.

higher resistance to corrosion 1. corrosion – the damaging of the surface of a metal by chemical action. It happens, for example, when a metal is exposed to damp air, or to salt water. Oxides form on the surface of the metal. When this happens on iron, we say the iron has become 'rusty'. Often this oxide eats deeper into the metal and damages it seriously. In aluminium, the oxide on the surface does not eat into the metal, but protects it from damage by chemical action. 2. higher resistance – aluminium has a higher resistance than iron to corrosion. It **resists,** or 'stands up against' the kind of damage you have just read about.

film a very thin sheet.

aircraft production the production of aeroplanes.

manufacturers the directors of factories where things are produced, or 'manufactured'.

overhead cables cables held up in the air by telegraph poles. Opposite – underground cables.

sagged hung down in a curve.

a core goes through the centre along the length of something.

cooking utensils contain food which is being cooked.

the fuselage the body of an aeroplane.

seaplane an aeroplane which can come down on the sea.

pure aluminium aluminium that has not been mixed with another metal to form an alloy.

three-ply metal has three layers. (We also talk about 'ply-wood' made of thin layers of wood stuck together.)

exposed it to left it open to.

Exercises

A Use these words to complete the following sentences: age hardening, bauxite, density, duralumin, foil, good conductor, heat, higher resistance, light, pure.

1 .... is an ore which contains aluminium.

2 The .... of aluminium is about a third of that of steel or brass.

3 Aluminium is useful for making aeroplanes because it is a .... metal.

4 Aluminium is useful for telephone cables because it is a .... of electricity.

5 Cooking utensils are made of aluminium because it is a good conductor of ....

6 Metal that has been rolled into very thin sheets and is used for wrapping food is called . . . .

7 Dr Alfred Wilm first made the aluminium alloy called . . . .

8 Duralumin becomes stronger through . . . .

9 Pure aluminium has a . . . . to corrosion than duralumin.

10 Thin layers of . . . . aluminium are added to each side of the duralumin which is used in aircraft production.

B Answer these questions.

1 Where was bauxite first discovered?
2 Name three parts of the world where bauxite is now mined.
3 Give three reasons why aluminium is a very useful metal.
4 How can aluminium be made stronger?
5 What protects aluminium from corrosion?
6 Why is aluminium useful for making aeroplanes?
7 Why is aluminium used instead of copper in overhead telegraph and telephone cables?
8 How did men make overhead aluminium cables firmer?
9 What metals are used to make duralumin?
10 Why is it important that the metal used to make seaplanes should resist corrosion?
11 How did men test the 'three-ply' metal to see if it resisted corrosion?

C Write sentences using each of the following words or phrases:

1. precious metal 2. corrosion 3. melting point
4. production 5. brittle

Petroleum

1. Introduction

Petroleum has many uses and it is used in many different forms. The synthetic detergent you use for washing your clothes doesn't look like the petrol you put into your car engine, but both the powder and the petrol were **originally crude oil.** Here are four facts that are usually true about oil: it is a liquid; it is not generally found in a form in which it can be **immediately used;** it is not generally available in the places where it is mainly wanted; and it is very difficult to find at all.

There are exceptions to all these facts. For example, crude oil found in tar sands in Alberta, Canada, is almost solid. Some crude oil found in Indonesia can be immediately used as **fuel.** In Holland, oil is found close to **a main area of demand.** The oil which helps to form the Pitch Lake in Trinidad is **obvious.**

But usually great effort and expense are needed to find oil in **commercial quantities** in areas often thousands of miles away from where most of it is going to be used. When it is found it is in a form that cannot be used immediately but has to be **refined.** It has to be transported and stored very carefully. There are four main parts of the petroleum industry:

1. EXPLORATION for crude oil and PRODUCING it when it is found.
2. REFINING the crude oil into the different forms in which it is used.
3. TRANSPORTING the crude oil to the **refineries,** and transporting the **refined products** from the refineries to where they are needed.
4. SELLING these products to those who need them.

Exploration for crude oil is going on all the time, but it is a **risky** and expensive business. Geologists find places where the rocks are formed in such a way that oil may be stored between them. To find out for certain whether oil is there it is necessary to drill a well. When a well is drilled it is very often found that there is, in fact, no oil in this particular place. We say that the well is 'dry'. In 1920, 33,911 wells were drilled for oil in the United States. Of these 7,364 were dry. In 1956, 57,111 wells were drilled, and 21,838 were dry. The proportion of dry wells has **increased considerably.** This is because the oilfields which can be found fairly easily and cheaply have been discovered a long time ago. The proportion of failures when **wild-cat wells** are drilled is much higher than in the total just given. In the United States in 1956 nearly nine out of ten wild-cat wells were found to be dry.

Exploration for oil is very expensive because likely areas are usually a long way from cities and the cost of transporting men and equipment and establishing places for them to live in is high, and the drilling equipment **costs a great deal.** In Ecuador, for example, two oil companies co-operated in a search for oil which lasted for six years. In 1950 the search was abandoned. No oil had been found. The search cost £9 million. In Colombia another company started a search for oil in 1916 and had spent $21 million by the time they first found oil in 1933. In the next six years they spent another $39 million before the first oil from this area was **brought on to the market** in 1939. In the Zubair **oilfields** in Iraq £18 million was spent in six years before oil was first produced in 1951.

Adapted from *The Petroleum Handbook* (Shell)

Explanation of words in bold type

originally at first.
crude oil oil as it is found under the ground or on the surface. It is not pure. It contains impurities.
immediately used used straight away.
fuel anything that is burnt to give heat or power.
a main area of demand a place where something is needed.

obvious it can easily be seen.

in commercial quantities Oil companies only try to get oil out of the ground if they find enough in one place to sell for more than the high cost of getting it out of the ground. If they find enough oil to sell for more than this high cost, we say they have found oil in **commercial quantities.**

refined Men take the impurities out of crude oil to refine it. This is done in a **refinery**. When the crude oil has been refined it has been turned into **refined products.** (These are **produced** in the refinery.) The petrol you use to drive your car engine, for example, is a refined product of crude oil.

risky risky business is a business where there is a danger of losing money.

increased considerably grown much larger.

wildcat wells wells that are drilled in areas where no oil has been discovered before. These are really 'exploration' wells.

costs a great deal costs a lot of money.

brought on to the market sent to places where it is sold.

oilfields areas where oil is found.

Exercises

A Use these words to complete the following sentences: crude, drill, dry, geologists, increased, Pitch, refined, refinery, solid, wildcat.

1 Petrol used to drive car engines was . . . . oil at first.
2 In tar sands in Alberta crude oil is found that is almost . . . .
3 In Trinidad oil is found in the . . . . Lake.
4 Before crude oil can be used it must be . . . .
5 Crude oil is refined in a . . . .
6 . . . . study rocks to find out which may be storing oil.
7 It is necessary to . . . . a well to find out for certain whether there is oil under the ground.
8 A . . . . well is a well that is drilled in a place where no oil is found.

9 An exploration well, drilled in an area where no oil has been discovered before, is called a . . . . well.
10 In recent years the proportion of dry wells drilled has . . . .

B Answer these questions.

1 What is synthetic detergent used for?
2 Give *two* facts that are usually true about oil.
3 Where is oil found close to a main area where it is needed?
4 What usually has to be done to the oil taken out of the ground before it can be used?
5 How do geologists help in the exploration for oil?
6 How do men find out for certain whether there is oil in a particular place?
7 Why has the proportion of dry wells increased?
8 What proportion of wildcat wells were found to be dry in the United States in 1956?
9 Give two reasons why the cost of exploring for oil is very high.
10 When was oil first produced from the Zubair oilfields?

2. Pipelines

Oil usually has to be transported great distances from the oilfield to where it is going to be used. There are two main methods of transporting the oil, by pipeline and by **tanker.** Water transport is usually much cheaper, and tankers are used wherever possible.

Pipelines are used to take the crude oil from the oilfields to the seaports or to refineries, and to take the refined oil from the refineries to where it is going to be used. It is very expensive to **install** pipelines. A pipeline across the desert may cost **as much as £80,000 per mile.** But pipelines help to save money by saving tankers from having to make long sea **voyages.** For example, oil exported from Iraq to Europe **used to be transported** by tankers through the Gulf, the Arabian Sea, the Red Sea, and the Suez Canal, into the Mediterranean Sea. In 1934 a pipeline was installed from the Kirkuk oilfield in Iraq to the Lebanese port of Tripoli. In 1952 another pipeline from Kirkuk was installed, leading to the Syrian port of

Banias. Tankers now have only short voyages from these Mediterranean ports to Europe.

The first oil pipeline was installed in 1865 in Pennsylvania, USA. It carried crude oil from an oilfield to a railway station. Soon after it was installed it was broken up by **wagon drivers. Previously,** they had carried the oil and they were angry because they had lost their jobs. This first oil pipeline was only 6 miles long and only 2 in. in diameter. The longest oil pipeline today is in Canada. It is 1,765 miles long, and its diameter is 30 in. in parts. Parts of the Kirkuk–Banias pipeline are 32 in. in diameter.

The first pipeline was made of lengths of iron pipes screwed together. In 1900 steel was used for making pipelines. As steel is stronger than iron, men were able to make pipelines with larger diameters. Later, men **welded** the pieces of steel pipe together. This made **the joints** stronger.

Pipelines are usually **constructed** by a number of **gangs,** each working on a separate section of the line. Some lines are laid on the surface, especially in **deserted country,** but the majority of pipelines are buried. On agricultural land pipelines must be buried deep enough so that they don't interfere with farming. In cold areas pipelines must be buried beneath the **frost line.** All buried lines **are coated** with a protective covering to save them from corrosion. This covering usually consists of a layer of **bitumen,** with a layer of asbestos felt or glass fibre on top, and a final layer of bitumen on top of that. Where pipelines cross **swamps,** rivers, or lakes they may be **encased in concrete.** This both protects them and gives them extra weight to stop them floating up when they are empty. Sometimes there may be corrosion inside the pipeline. From time to time a **scraper** is passed through the pipeline to remove this corrosion.

Sometimes a pipeline travels downwards and the force of gravity alone is enough to move the oil through it. But usually the oil must be pumped through the pipeline. Pumping stations are placed where they are needed along the line.

Adapted from *The Petroleum Handbook*
(Shell)

Explanation of words in bold type

tanker	a ship specially designed for carrying oil.
to install	to put in place for the first time.
as much as	not more than, though it may be less than.
per mile	for each mile.
voyages	journeys made by sea are called voyages.
used to be transported	it was transported in the past. Now it is no longer transported. **I used to be** a schoolboy. Now I am a student.
wagon drivers	men who drove the carts (wagons) pulled by horses.
previously	before this.
welded	to weld is to join two pieces of metal together by heating the parts where they join until they melt and then pressing them together.
the joints	are the parts where the pieces of pipe are joined together.
constructed	made, put together.
gangs	groups of workers.
deserted country	is country where no people live.
the frost line	In very cold weather the ground may freeze down to a certain depth. We can imagine that there is a line through the earth at this depth. This is called **the frost line.**
are coated	are given a covering.
bitumen	a solid or semi-solid form of oil.
swamps	areas of land mostly covered by shallow water.
encased in concrete	fitted into concrete, like clothes being fitted into a case.
a scraper	scrapes the inside of a pipeline. It has sharp edges to cut off the corrosion.

Exercises

A Use these words to complete the following sentences: diameter, interfere, iron, oilfield, pipelines, protective, scraper, screwed, steel, tankers, voyages, welded.

1 Ships used to carry oil across the sea are called . . . .

2 Crude oil often travels from oilfields to seaports in . . . .

3 A pipeline leads from the Kirkuk . . . . to two Mediterranean ports.
4. Pipelines save tankers from having to make long . . . .
5 A pipeline constructed in 1865 was only 2 in. in . . . .
6 . . . . pipes were used for the first oil pipelines.
7 Nowadays . . . . is used to make pipelines.
8 The first pipelines were . . . . together.
9 Nowadays pipelines are . . . . together.
10 Pipelines must not . . . . with farming land.
11 When pipelines are buried they are given a . . . . covering.
12 Corrosion inside a pipeline is removed by a . . . .

B Answer these questions.

1 What are the two main methods of transporting oil?
2 Which method is cheaper?
3 How do pipelines help to save money?
4 To which Mediterranean ports do the oil pipelines from Kirkuk lead?
5 What was the first oil pipeline used for?
6 Why did wagon drivers break up this pipeline?
7 How long is the longest oil pipeline in the world?
8 Why is steel now used instead of iron for making pipelines?
9 How did men make the joints in pipelines stronger?
10 Give two things that are used to protect pipelines buried under the ground.
11 Why are pipelines passing through water sometimes encased in concrete?
12. What force makes oil pass through pipelines that travel downwards?
13 How is oil usually forced through pipelines?

C Use each of the following words in a sentence:

1. oilfield 2. refine 3. pipeline 4. tanker 5. exploration

3. Uses of Petroleum

When petroleum is first taken from the ground it is not very useful. After it has been treated in refineries it has many different uses. Most of the world's oil supply is used as some kind of fuel. The fuel may be used in different kinds of engines to give power, or to give heat in factories and in homes. Some of the world's oil is used for **lubricating** machinery, and some of it, in solid or **semi-solid** form, called bitumen, is used in making roads. There are also many other uses for **oil products.**

For many years men used solid fuels, such as wood and coal, to provide power and heating. Oil is a liquid fuel. It is more useful than solid fuel for several purposes. When men discovered how to use liquid fuel it was possible to invent motor vehicles and aëroplanes. All fuels need air for burning. Liquid fuels easily **evaporate** to form vapour, which mixes well with the air and **therefore** burns quickly. Solid fuels cannot easily be **dispersed** in the air, so it is more difficult to burn them quickly. Liquid fuels burn up almost **entirely,** but solid fuels only burn in part, leaving behind some ash. Also, liquid fuels are more easily stored and handled than solid fuels.

A liquid fuel which easily evaporates and turns into vapour is called 'volatile'. The most volatile oil fuel is gasoline, and this is used in most motor engines. (In Britain this fuel is called 'petrol'.) When oil was first being refined a hundred years ago the **internal combustion engine** used in motor cars had not been invented so very little gasoline was produced. But great quantities of another oil fuel, kerosene, were produced, for burning in lamps. Gasoline is too **inflammable** for this purpose.

Kerosene was the most important product for fifty years. It is a colourless liquid and is less volatile than gasoline. It was first used as a lamp oil, and it is still used for this purpose. It is now also used for cooking and heating.

Diesel fuels are less volatile than kerosene. They are used to drive diesel engines, either **stationary engines** or **automotive engines.** Many heavy lorries use diesel, and it is also used for **marine engines.**

Lubricating oils are used to make sure that machinery runs smoothly. Even the most carefully finished metal surface is not truly flat. It is covered with tiny **projections** and **depressions.** When two dry surfaces are placed against each other they may lock together, and they will resist our effort to slide one surface over the other. This resistance is called friction. When lubricating oil is put on the surfaces it fills up the depressions and covers the projections. A film of oil is formed on the surfaces. Now we can easily slide one surface over the other. This is just one way in which lubricating oil helps machinery to run smoothly.

After crude oil is refined to make fuel oils or lubricating oils, a solid or semi-solid product is left. We call this bitumen. It is useful in road making. Bitumen is also found in natural forms, sometimes mixed with mineral matter. Five thousand years ago people in Iraq used it. They found it **seeping through** the surface of the earth. There is a lake in Trinidad which contains a huge supply of bitumen mixed with mineral matter.

Adapted from *The Petroleum Handbook* (Shell)

Explanation of words in bold type

lubricating covering something with oil to make it smooth.
semi-solid half solid.
oil products things made from oil.
evaporate change into a vapour.
therefore because of this. He had not been able to sleep all night. **Therefore** he was too tired to work well.
dispersed spread out.
entirely completely. Is the paper entirely burnt up? No, a little piece of it is still not burnt.
internal-combustion engine an engine in which the fuel is burnt inside the cylinder. (Internal – inside; combustion – burning.)
inflammable can easily be made to burn.
stationary engines work standing in one place.
automotive engines work moving from one place to another. (They are in **motion.**)
marine engines drive anything that travels on the water.
projections pieces that stand up from the surface.
depressions holes in the surface.
seeping through coming through slowly in small quantities.

Exercises

A Use these words to complete the following sentences:
air, ash, bitumen, depressions, diesel, evaporate, gasoline, heat, kerosene, liquid, natural, power, projections, refineries, solid.

1 Most oil must be treated in . . . . before it is useful.
2 Oil fuel gives . . . . to engines, and . . . . in factories and homes.
3 An oil product used for making roads is called . . . .
4 Wood and coke are . . . . fuels.
5 Oil is a . . . . fuel.
6 Liquid fuels burn quickly because they . . . . and mix well with the . . . .
7 Solid fuels leave behind some . . . . after they are burnt.
8 . . . . is the most volatile oil fuel.
9 More . . . . than gasoline was produced when oil was first refined.
10 . . . . fuel is often used in heavy lorries and in marine engines.
11 Lubrication oil fills up the . . . . and covers the . . . . which cover the surface of parts of machinery.
12 Five thousand years ago bitumen was found in Iraq in its . . . . form.

B Answer these questions.

1 What is most of the world's oil supply used for?
2 What did men use to provide power and heating before oil fuel was widely used?
3 Why does liquid fuel burn more quickly than solid fuel?
4 What kind of oil fuel is used in most motor-car engines?
5 What was kerosene first used for?
6 What is kerosene used for nowadays?
7 Why isn't gasoline used for burning in lamps?
8 Give two kinds of engines that can be driven by diesel oils.
9 Why may two carefully finished metal surfaces resist our efforts to slide them against each other?
10 Where can bitumen be found in its natural form?

C Use each of the following words and phrases in a sentence:

1. bitumen 2. fuel 3. crude oil 4. power
5. easily evaporates.

4. Exploration

Thousands of years ago men used petroleum for lamps. They found this petroleum seeping out of the ground, or collected it from the surface of ponds. Just over a hundred years ago men began drilling holes near the places where oil was seen seeping out of the ground. Later, **geologists** were asked to help to find oil under the ground. They said that there was likely to be oil in the **crests** of folds of rock layers. These folds are called anticlines. A great many successful holes were drilled where anticlines were found.

Areas where large amounts of oil are likely to be found are generally **low-lying.** They are along coastlines where the seas are **shallow** or in the **foothills** and lowlands beside mountain ranges. The very rich oilfields in the south of Texas and Louisiana are alongside the shallow sea of the Gulf of Mexico, and the oilfields around the shallow sea between Iran and Arabia are also very rich. The geologist studies not only the way in which layers of rock are folded but also what kinds of rock form these folds. Rocks in which oil is formed are called **sedimentary rocks,** but oil is actually found in cracks between other kinds of rocks which do not allow the oil to seep away.

Before a new area is explored from the ground it is photographed from the air. An aeroplane fitted with a special **wide-angle camera** flies over the area by strips, taking photographs each of which overlaps those next to it by more than 50%. After a careful study of these photographs a geological map is made. This map shows the main geological features which can be seen from the air. Geologists can then go directly to study the most interesting parts of the area.

The geologist then looks for places in the most interesting parts where he can study the formation of the rocks more clearly. He can do this in the banks of rivers, in mountain slopes, and in the sides of **cliffs.** He studies the kinds of rock and sometimes sends pieces of these rocks back to his headquarters where they are looked at even more closely. Sometimes when there are no river banks, mountain slopes, or cliffs, it is necessary to dig trenches or to drill holes so that the geologist can study the rocks.

In some areas where oil may be found, such as deserts, or beneath shallow waters, there is no evidence on the surface that the geologist can study. In these areas a **geophysical exploration** is made. One method used in a geophysical exploration is called the **seismic** method. Explosives are let off at the surface of the earth. These cause vibrations beneath the surface, and these vibrations are partly **reflected** to the surface from the points where layers of rock of different density meet. Instruments called seismometers **relay** these reflected vibrations to a recording instrument. The recording instrument shows how long it took for the vibrations to travel from the surface down to the reflecting point and back again. Scientists can learn from this the depth of the point where the layers of rock of different density meet.

When all the necessary geological and geophysical information about an area has been studied, exploration wells are drilled in the most likely part of the area. Sometimes the first wells give only more geological information. Sometimes enough oil shows to encourage the drilling of other wells near by. And sometimes men have the good fortune to discover rich oilfields.

Adapted from *The Petroleum Handbook* (Shell)

Explanation of words in bold type

geologists study the kind of rocks forming the earth's surface, and how these rocks are formed.

crests the tops. The top of a hill is called the **crest** of a hill.

low-lying ground that is low, without hills and mountains.

shallow not deep.

foothills the small hills beside higher hills or mountains.

sedimentary rocks are formed of different materials and the remains of living matter.

a wide-angle camera can take a photograph showing a large area.

cliffs sides of rocks standing straight up from the ground.

geophysical exploration is an exploration to find out the way certain forces behave in the earth. A geophysicist is a scientist who studies this.

seismic is a word formed from **seism**, which means an earthquake, a shaking in the earth.

reflected sent back.

relay send on.

Exercises

A Use these words to complete the following sentences: crests, depth, explosives, geological, geophysical, low-lying, sedimentary, seeping, surface, wide-angle.

1 At first men drilled holes where they saw oil . . . . out of the ground.

2 Geologists said oil was likely to be found in the . . . . of folds of rock layers.

3 Where large amounts of oil are found, the ground is generally . . . .

4 Oil is formed in . . . . rocks.

5 Photographs of new areas are taken from the air with . . . . cameras.

6 A . . . . map is made after the photographs have been studied.

7 Where there is no evidence on the surface for a geologist to study a . . . . exploration is made.

8 In the seismic method . . . . are let off at the . . . .

9 By the seismic method geophysicists can learn about the . . . . of different layers of rock.

B Answer these questions.

1 How did men collect oil thousands of years ago?

2 Give the names of *two* places where oil is found near shallow seas.

3 What does the geologist study to help in the exploration for oil?

4 How is a new area first studied?

5 What does the geological map made from the photographs show?

6 Give *two* kinds of places where a geologist can easily study the formation of the rocks.
7 In what areas are geophysical explorations made?
8 What does a seismometer do?
9 What does the recording instrument do?
10 What happens after all the necessary geological and geophysical information has been studied?

C Use each of the following words in a sentence:

1. shallow 2. explore 3. drill 4. vibrations.

Ventilation

1. Introduction

WHAT VENTILATION DOES

Effective ventilation helps people to work well and to keep healthy. It is necessary to keep the atmosphere pure, and to control the temperature and movement of the air.

PURE AND IMPURE AIR

The composition of pure air is approximately 21% oxygen and 79% nitrogen. Small portions of carbon dioxide, argon, neon, and helium are also present. In large towns the carbon dioxide content may be slightly higher than in country districts, but **it rarely exceeds 0.04%.**

In **occupied buildings** the atmosphere is affected by the **bodily functions** of the **occupants.** Carbon dioxide and **water vapour** are given off in the air that is **exhaled** from the lungs. Bacteria are **discharged** by people breathing, sneezing, or coughing. The atmosphere may be **contaminated** by people smoking, by flames, by fumes, gases, or dust. Good ventilation helps to **reduce** all these impurities. For example, when the carbon dioxide content in the atmosphere is increased, the oxygen content is decreased by a similar amount, but in a well-ventilated building the carbon dioxide content will not be more than 1%.

BODY HEAT

The human body **continuously** produces heat, which it **derives from** eating food. When a man is working his body produces more heat than when he is at rest. But the body cannot store heat. The inner temperature of the body must **remain constant** at about 37° C. People become ill if their temperature rises much above this.

The body must **dissipate** heat as fast as it produces it. In cold surroundings the loss of heat is quick, but in a warm building the body may gain heat from its surroundings. It must dissipate not only the heat it produces but also the heat it gains.

If somebody's body cannot dissipate heat quickly enough he may become ill. Certainly he will not be able to work well. Body temperature is regulated by three processes: radiation, convection, and evaporation.

RADIATION

The body transmits heat by radiation and receives heat by the same process. In a room with cool walls the body will lose more heat by radiation than it receives. If the walls are hotter than the skin the body gains heat by radiation.

CONVECTION

The body loses heat by convection when the air is cooler than the skin. **Air motion accelerates** this process of heat loss. If the air was **absolutely still** and the body was still, too, the air around the body would be warmed to about the same temperature as the skin. Then the body would not be able to dissipate any more heat. But when the warmed air around the body is carried away by the motion of the air the rate of convection is accelerated. Then the skin temperature is lowered. That is why a fan which is moving the air in a room helps us to feel cooler, **so long as** the air temperature is not warmer than that of the skin. But if the air in a room is warmer than the skin, the convection process increases body heat.

EVAPORATION

If the heat taken from the body by radiation and convection is not sufficient to keep it at normal temperature, the body begins to perspire. Small drops of moisture are formed on the outer surface of the skin. When this moisture evaporates in the atmosphere the body loses heat. But sometimes the atmosphere surrounding the body cannot easily **absorb** more moisture. This is true when the atmosphere is still and **humid.** If the moisture on the surface of the skin is not absorbed, the body cannot continue to perspire and it cannot lose heat in this way. But if the air is made to move, and fresh air is brought close to the body, heat loss through perspiration may be possible. (Unfortunately this is not true in districts where the atmosphere is very humid.)

To sum up, air motion may accelerate body heat loss both by convection and by assisting the body to perspire freely. But it is not good for the body to perspire too much, chiefly because it may be weakened by the loss of salt, carried away in the evaporating drops of moisture. A good ventilation system provides enough air movement to accelerate heat loss from the body by convection alone, and does not allow the body to perspire freely.

Adapted from *Woods Practical Guide to Fan Engineering*, edited by W. C. Osborne and C. G. Turner (Woods)

Explanation of words in bold type

it rarely exceeds 0.04%	it is not often more than 0.04%.
occupied buildings	buildings in which there are people.
bodily functions	the work done by the human body.
occupants	the people in a building.
water vapour	water in a state of gas.
exhaled	breathed out. Opposite: inhaled – breathed in.
discharged	sent out.
contaminated	made impure (not pure).
to reduce	to make something less.
continuously	all the time.
derives from	receives from.
remain constant	stay exactly the same.
dissipate	send away.
air motion	the movement of the air.
accelerates	makes faster, speeds up. You press the **accelerator** in a car to make the car go faster.
absolutely still	not moving at all.
so long as	1. You will pass your examinations **so long as** you study all the year. 2. You will be safe on the roads **so long as** you drive carefully.

absorb take in. If you spill some ink on your examination paper, use blotting paper. It will absorb the ink.

humid atmosphere has a lot of moisture in it. It cannot absorb much more moisture from the body when it perspires.

Exercises

A Use these words to complete the following sentences: absorbed, carbon dioxide, constant, contaminated, convection, dissipate, increase, moisture, oxygen, well ventilated.

1 Pure air is composed of 21% . . . . and 79% nitrogen.
2 Air exhaled from people's lungs gives off . . . . and water vapour.
3 When there are bacteria, fumes, gases, or dust in the air we say that the air is . . . .
4 The carbon dioxide content in the air will not be more than 1% in a building that is . . . .
5 The inner body temperature of healthy people stays . . . . at about 37° C.
6 To remain healthy the body must . . . . heat as fast as it produces it.
7 . . . . is the process by which the temperature of the body is regulated by the action of the air on the skin.
8 The convection process will . . . . body heat if the air around the body is warmer than the skin.
9 When the body perspires, small drops of . . . . are formed on the outer surface of the skin.
10 The body cannot continue to perspire if the moisture on the skin is not . . . . into the atmosphere.

B Answer these questions.

1 How is bacteria discharged in occupied buildings?
2 Name three things that may contaminate the air in a building.
3 What happens when the carbon dioxide content in the atmosphere is increased?
4 From where does the human body derive the heat which it produces?

5 What is the normal inner temperature of the human body?
6 What are the three processes which regulate the body's temperature?
7 How may air motion help the process of convection to make the body lose heat?
8 What happens to the human body if the processes of radiation and convection are not sufficient to keep the body at normal temperature through heat loss?
9 How may air motion help the process of evaporation to make the body lose heat?
10 Why may the body become weakened if it perspires too much?

C Use each of the following words in a sentence:

1. impure 2. motion 3. decrease 4. accelerate 5. perspire

2. Estimating Ventilation Requirements

Some people think that ventilation is the same thing as air conditioning. Air conditioning means that the air supply to an occupied building is brought to a required condition by cleaning, cooling, heating, drying, or **humidifying.** We say that the air is 'conditioned' by these processes. Any kind of atmosphere can be produced like this, but it costs a lot of money. Ventilation is concerned with the supply of fresh air, the removal of **contaminants** and heat, and the control of air motion for cooling and **freshening.**

The essential requirement in ventilation is to replace contaminated and overheated air with fresh **outdoor air,** and to prevent people **feeling discomfort** from **humidity.** To estimate the amount of ventilation required **the following factors have to be taken into account:**

1. The size of the room or building.
2. The number of occupants, and **their activities.**
3. Heat gain from equipment and from **solar radiation.**
4. The humidity of the atmosphere.
5. The air temperature outside the building.

The ventilation should be designed to **deal with** the hottest conditions that there are **likely** to be for any length of time.

Air needs to be changed in buildings a certain number of times every hour. In some buildings air needs to be changed more **frequently** than in others. For example, the air in a house where one or two people are living need not be changed as often as the air in a cinema where there are a lot of people (especially if they are allowed to smoke), or in a paint shop where many people are working. Both cigarette smoke and paint fumes contaminate the atmosphere. There are two methods of estimating the amount of fresh air that is needed in a building. One method is **based upon** the interior size of the building and the other method is based upon the number of people who are using the interior of the building at the same time.

When we estimate how much ventilation is required to reduce the temperature in a building we have to think about heat-gain factors; that is, the things which make the temperature increase. Heat gain may come from the following **sources**:

1. Body heat given off by the occupants of a building.
2. Heat given off by electrical apparatus and by machinery.
3. Heat transmitted through walls and roof by the sun's rays.

Even when a man's body is at rest it gives off a certain amount of heat. If the man is working the body heat given off is increased. When a number of men are working together in one room the amount of heat they give off is **considerable.**

When electrical motors and other electrical apparatus are being used they become warm and give off heat into the atmosphere. As the motors are turning they may be helping to send out warmth. Electrical apparatus, such as lighting, also gives off heat.

The interior temperature of a building may be raised by solar heat transmitted through windows, walls, and roofs. In strong sunshine the **external** surfaces on which radiation falls may become as much as 15·5° C. hotter than the atmosphere. Part of this heat will be given off into the air, but part will **penetrate** into the walls.

Building surfaces of dark colours absorb more heat than those of light colours. Some building surfaces allow more heat to penetrate through them than others. Heat **penetration** depends on how thick the surfaces are, and what **substance** they are made of. Solar heat passes almost immediately through glass, and very quickly through corrugated iron. Special substances may be used **to insulate walls.** It will take a long time for solar heat to pass through a wall that is very thick and well insulated. If the heat of the day has finished before solar heat has had time to penetrate **such a wall,** this heat can pass out of the building again into the cool of the night.

But in most buildings in hot climates it is necessary to estimate the amount of heat gain within the building from solar radiation. It must be remembered that radiation affects only one or two walls of a rectangular building at the same time. In the earlier part of the morning solar radiation falls on to east walls, from then until early afternoon on to south walls (or north walls in southern latitudes), and after that on to west walls. Estimates of the amount of ventilation required to remove heat gain from solar radiation should be based upon the wall which transmits the greatest amount of solar heat at the hottest time of the day.

Adapted from *Woods Practical Guide to Fan Engineering*, edited by W. C. Osborne and C. G. Turner (Woods)

Explanation of words in bold type

estimating ventilation requirements trying to find out exactly how much ventilation is required.
humidifying making the air humid by putting moisture into it.
contaminants impurities. This air is contaminated. It contains contaminants.
freshening making fresh.
outdoor air the air outside buildings.
feeling discomfort not feeling comfortable.
humidity a humid atmosphere.

the following factors have to be taken into account	You have to think about the following points.
their activities	what they are doing.
solar radiation	the heat sent out by the sun.
deal with	You know how to repair cars. You can deal with the work in the garage. I know how to type letters. I can **deal with** the work in the office.
likely	It is now August. The weather conditions are **likely** to be hot.
frequently	often.
based upon	The price of fruit in the market is **based upon** the cost of growing and transporting it.
sources	are the beginnings of things. The source of a river is where it comes out of the ground.
considerable	quite a lot.
external	outside.
penetrate	go through.
penetration	the process of going through something.
substance	material; here a building material. Concrete is a **substance** made of cement, sand, and small stones.
to insulate walls	is to make them proof against certain effects, for example, of heat, or of sound.
such a wall	a wall like the one just written about.

Exercises

A Use these words to complete the following sentences: conditioned, contaminate, discomfort, frequently, freshen, interior, light, penetrate, transmit, working.

1 When air is cleaned, cooled, heated, dried, or humidified by a special process, we say that it is ....
2 Ventilation controls air motion in order to cool or .... the atmosphere.
3 If the air is too humid, people feel ....
4 The air in a cinema containing many people needs to be changed more .... than the air in your room.
5 Paint fumes .... the atmosphere.
6 One method of estimating ventilation requirements is based upon the .... size of the building.

7 The sun's rays may . . . . heat through the walls and roof of a building.
8 The heat given off by a man's body is increased when he is . . . .
9 Substances of dark colours absorb more heat than those of . . . . colours.
10 It is difficult for heat to . . . . a thick and well-ventilated wall.

B Answer these questions.

1 Name three things that air conditioning may do to the air.
2 When estimating ventilation requirements, what are two things about the people in a building that we have to take into account?
3 Why does the air in a paint shop have to be changed frequently?
4 What are 'heat-gain factors'?
5 Name two sources of heat gain in a building.
6 Name a kind of electrical apparatus that gives off heat.
7 Name two kinds of building surfaces through which solar heat passes very quickly.
8 What happens to the solar heat in a wall if the air outside the building becomes cooler at night?
9 How many walls of a rectangular building are affected by solar heat at the same time?
10 Which wall of buildings receives solar radiation at midday in southern latitudes?

C Use each of the following words in a sentence:

1. frequently 2. penetrate 3. contaminate 4. humid 5. insulate

3. Fan Ventilation Systems

Controlled ventilation requires the use of fans. **That applies to** most industrial and commercial buildings, especially where the number of occupants is large, or where heat or steam must be removed. Natural ventilation can only be satisfactory if outside weather conditions are **ideal.** A natural supply of fresh air depends on the wind and on the difference between the inside and outside temperatures. Fresh air will not flow naturally into a building unless a **breeze** is blowing in the right direction. Warm, contaminated air will not flow out of a building unless the outside air is cooler than the inside air. Controlled ventilation is most needed in hot, **windless weather,** and it is also necessary to control the dust or **odours** in the **incoming air.**

There are three systems of fan ventilation:

1. **Extraction** of air.
2. Supply of air.
3. A combination of extraction and supply.

THE EXTRACTION SYSTEM

The extraction system is the most widely used. The interior air is extracted from the building and this causes fresh outside air to enter. Propeller fans **exhaust** air through holes in the wall.

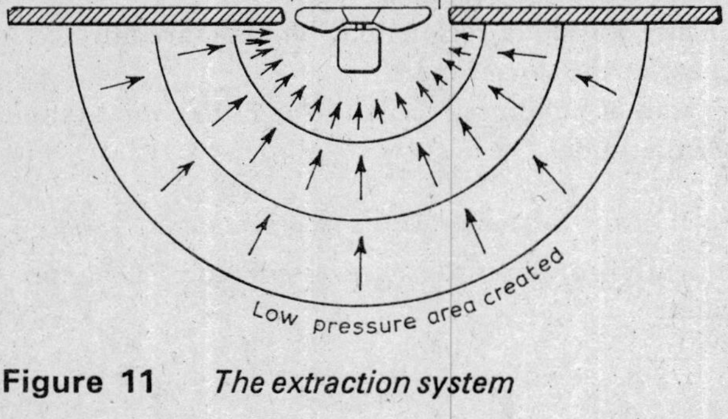

Figure 11 *The extraction system*

When this air has been exhausted an area of low pressure is created near it. Other air flows into that low-pressure area and creates a wider area of low pressure behind it. Then fresh air from outside flows into this area through any **available openings** in the walls. (See figure 11.)

The placing of such openings (or 'inlets') is important. See figure 12. There is only one **inlet** into the room and the fresh air is flowing across the room towards the **exhaust point.** The

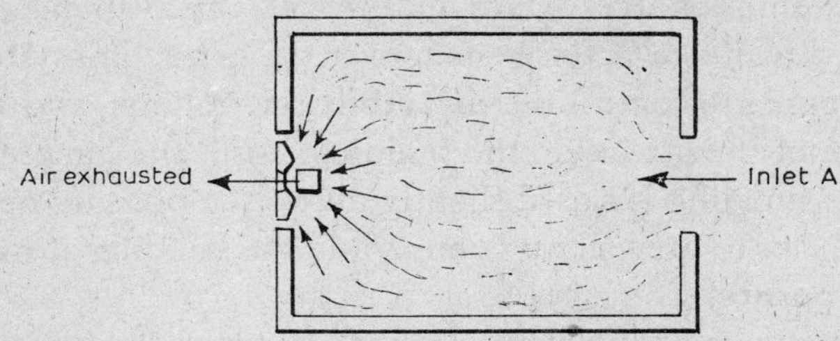

Figure 12 *A well ventilated room*

room is being well ventilated. Now look at figure 13. Here there are two inlets into the room, but the fresh air entering at inlet *B* is flowing straight to the exhaust point. It is being drawn out of the room before it can ventilate it very much. Also,

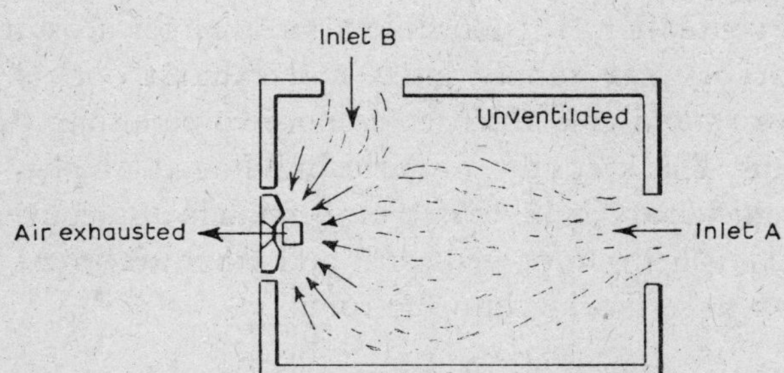

Figure 13 *Room not so effectively ventilated*

because the extraction fan is dealing with this air from inlet *B*, it is not drawing the air across the room from inlet *A* so effectively.

Another point to be considered is that fresh air becomes contaminated as it passes across an occupied room. If the distance between the inlet and outlet points is too long, people nearest to the outlet point will be breathing air that has been contaminated too much.

THE AIR SUPPLY SYSTEM

A supply system is the opposite to an extraction system. Fresh air is blown into an occupied building by fans and the contaminated interior air will flow out through any available opening. The advantage of this system over the extraction system is that it is easier to control the **distribution**, volume, and speed of motion of the air inside the building. Also, the air pressure inside the building is raised slightly above the pressure outside and this helps to prevent air coming into the building at **inconvenient points.**

In the supply system it is necessary to place the air outlets very carefully so that the fresh air is well distributed **throughout** the building. It is necessary to consider the weather conditions outside the building. For example, wind blowing strongly against a wall containing outlet openings may stop the **outflow** of contaminated air.

COMBINED SUPPLY AND EXTRACTION SYSTEM

This system, using both extract and supply fans, gives the fullest control of ventilation. It is possible to arrange for a controlled flow of air between supply inlets and exhaust outlets. The **input fans** should give about 20% more volume than the extraction fans. This keeps the air in the building at a higher pressure than outside and helps to stop air coming in at inconvenient points. Also, it helps to prevent dust and other **airborne contaminants** from flowing into the room.

Adapted from *Woods Practical Guide to Fan Engineering*, edited by W. C. Osborne and C. G. Turner (Woods)

Explanation of words in bold type

That applies to	that is true about. All traffic should keep to the right. That rule **applies to** horses and bicycles as well as to cars.
ideal	exactly what is wanted.
breeze	wind blowing in one direction.
windless weather	weather when there is no wind. (Wind**less** is formed like wire**less.**)

odours smells.
incoming air air coming in. Opposite – outgoing air.
extraction taking out.
exhaust blow out.
available openings holes that are there to be used.
The placing choosing the position.
an inlet lets the air into the room. Opposite – an outlet.
the exhaust point where the air is exhausted (blown out of the room).
another point to be considered something else to be thought about or to be taken into account.
distribution sending around. Milk is **distributed** around the city in lorries fitted with refrigerators.
inconvenient points places where air is not wanted.
throughout all the way through.
outflow flowing out. Opposite – inflow.
input fans fans which put air into the room.
airborne contaminants contaminants (or impurities) carried in the air.

Exercises

A Use these words to complete the following sentences: combination, contaminated, direction, exhausted, extraction, flows, low pressure, occupants, supply, volume.

1 Controlled ventilation is usually needed in buildings where the number of . . . . is large.
2 A building can be ventilated naturally if a breeze is blowing in the right . . . .
3 The most widely used controlled ventilation system is the . . . . system.
4 In the extraction system air is . . . . through holes in the wall.
5 An area of . . . . is created in the area from which air has been extracted.
6 If the placing is well chosen fresh air . . . . across the room from the inlet to the exhaust point.
7 Fresh air passing across an occupied room becomes . . . .

8 The opposite system to the extraction system is called the .... system.
9 The fullest control of ventilation is given by a .... of the supply and extraction systems.
10 In the combined supply and extraction system the .... of air supplied by the input fans is about 20% more than that exhausted by the extraction fans.

B Answer these questions.

1 Name two things that may have to be removed from the air in an occupied building by controlled ventilation.
2 In what weather conditions is controlled ventilation most needed?
3 What are the three systems of fan ventilation?
4 Why is the inlet point in the room shown in figure 12 well placed?
5 Why should the inlet and outlet points be not too far apart?
6 Name three things that can be controlled more easily by using the supply system than by using the extraction system.
7 Why should the air pressure inside the building be slightly higher than the air pressure outside?
8 What may stop contaminated air from flowing out of a building through the outlet openings?
9 What should be the difference between the volume of air blown into a building and that exhausted when both input and output fans are used?
10 What helps to stop airborne contaminants from flowing into a room?

C Use each of the following words in a sentence:

1. controlled 2. exhaust 3. inlet 4. occupied 5. distribute

4. Fans

A fan is a machine which **propels** air **continuously** by **aerodynamic action.** There are three main types of fans: centrifugal, propeller, and axial flow.

CENTRIFUGAL FANS

The **centrifugal** fan has an **impeller** wheel which **rotates** in a casing shaped like a **scroll.** (See figure 14.) The impeller

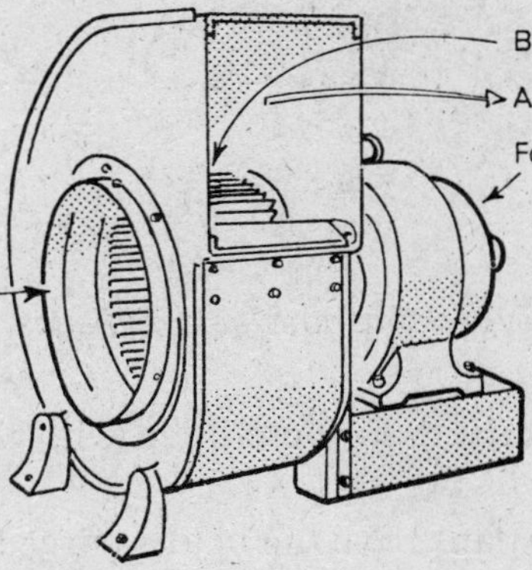

Figure 14 *Centrifugal fan*

wheel has a number of blades around its **periphery.** The air comes into the fan at the **axis** of the wheel and is exhausted at right angles to it. When the impeller wheel rotates, the blades at its periphery throw off air **centrifugally** in a direction following the rotation. The air thrown off into the scroll is forced out of the outlet as more and more air leaves the blades. At the same time air is sucked into the inlet to replace the exhausted air. The air enters **axially,** turns at right angles through the blades, and is discharged **radially.**

The blades of a centrifugal fan are usually one of three main types: straight radial blades, forward curve blades, or backward curve blades. (See figure 15.) The straight radial-bladed fans are generally used to deal with **moderate** pressures of air. They are not very efficient in conditions where air pressure is high. One good feature about this type is that the blades do not pick up matter carried in the air stream, as curved blades may do. A centrifugal fan with straight radial blades can be used in **dust-laden air.**

Forward curve fans have the blades curved with the concave side facing the direction of rotation. Usually a large number of blades are attached to each wheel and the fans are sometimes known as 'multiblade' fans. Fans with these blades throw off air

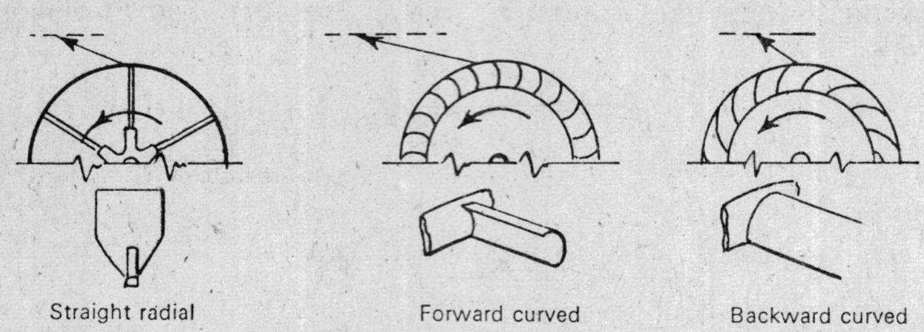

Figure 15 *Types of centrifugal fan blades*

at a greater **velocity** than those fitted with the other types of blades.

Backward curve fans have the blades curved with the convex side facing the direction of rotation. The air flows more smoothly because there is less shock when the air entering through the axis comes up against the blades. These fans have to be rotated at higher speeds than the other types of centrifugal fans.

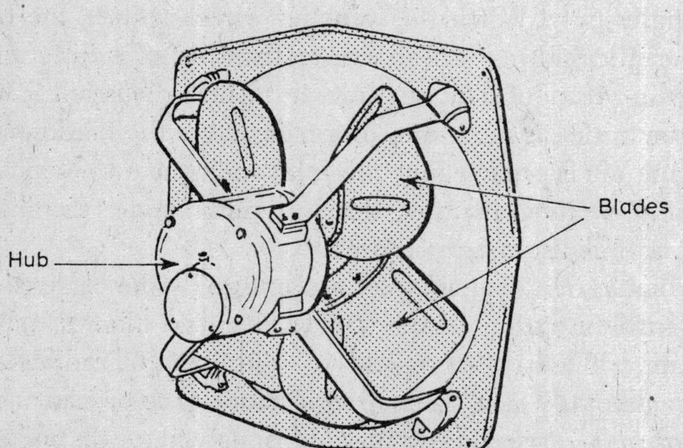

Figure 16 *Propeller fan with four blades*

PROPELLER FANS

These fans are widely used where resistance to air flow is low. Unlike the other types of fans, they are not usually used in a

duct system. Usually they move air through a hole in a wall or operate freely within a room. These fans have an impeller with two or more blades, usually made of sheet steel and set at an angle to the **hub.** They look something like a ship's propeller. (See figure 16.)

Air enters a propeller fan from all directions and is **discharged** mainly axially, but there is also some radial discharge. Propeller fans are not effective in conditions where air pressure is high because the air can flow back through the impeller.

AXIAL FLOW FANS

These fans have an impeller, or impellers, with specially shaped blades rotating inside a **cylindrical casing.** The air enters the fan axially and is discharged axially. (See figure 17.) This makes it more efficient than the centrifugal fan, in which the air has

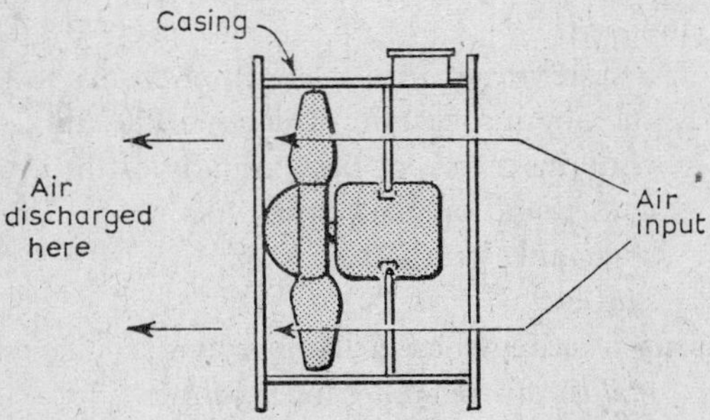

Figure 17 *Axial flow fan*

to pass through an angle of 90° before being discharged. Like the centrifugal fan, it is usually used inside air ducts but it can be fitted more easily into a duct because it does not need to be connected at 90°. Also an axial flow fan is smaller than a centrifugal fan designed to do the same duty. Because of their efficiency, **compactness,** and **simplicity of installation,** axial flow fans have become **increasingly used** in recent years.

Adapted from *Woods Practical Guide to Fan Engineering*, edited by W. C. Osborne and C. G. Turner (Woods)

Explanation of words in bold type

propels pushes out.
continuously all the time.
aerodynamic action the action of air in motion.
centrifugal moved by a force from the centre.
impeller something which sets something else in motion.
rotates goes round.
scroll a roll of paper, or a design looking like the end of a roll of paper.
periphery the circumference of the wheel.
axis the centre of the wheel. The wheel rotates around the axis.
centrifugally from a centre.
axially through the axis.
radially from the outer ends of the radii of a wheel.
moderate pressures pressures that are not very high, and not very low.
dust-laden air air that contains dust.
velocity speed.
duct system a system that uses a length of casing to lead the air in a certain direction. The duct contains a fan, or fans, usually of the centrifugal or axial flow type.
hub the centre of a wheel.
discharged sent out.
a cylindrical casing a casing shaped like a cylinder; like a length of iron pipe, for example.
compactness smallness.
simplicity of installation they are easy to install.
increasingly used used more and more.

Exercises

A Use these words to complete the following sentences: axially, axis, backward curve, centrifugally, concave, cylindrical, duct, high, right angles, rotates.

1 The impeller wheel in a centrifugal fan . . . . in a casing shaped like a scroll.
2 Air enters a centrifugal fan at the . . . . of the wheel.
3 The blades of a centrifugal fan discharge air . . . .

4 The air which enters a centrifugal fan turns at .... when the blades catch it.
5 Straight radial-bladed centrifugal fans are not very efficient where air pressure is ....
6 The .... side of the blades face the direction of rotation in forward curve fans.
7 The convex side of the blades face the direction of rotation in .... fans.
8 Air from axial flow fans is discharged ....
9 The blades of axial flow fans rotate inside a .... casing.
10 Both centrifugal and axial flow type fans are often used in a .... system of ventilation.

B Answer these questions.

1 What are the three main types of fans?
2 What shape is the casing inside which the wheel of a centrifugal fan rotates?
3 Where does air enter a centrifugal fan?
4 What are the three main types of blades used in centrifugal fans?
5 What is a good feature of centrifugal fans fitted with straight radial blades?
6 Which type of blade throws off air at the greatest velocity?
7 Which type of centrifugal fan has to be rotated at the highest velocity?
8 Which type of fan is not usually used in a duct system?
9 Why can axial flow fans be more easily fitted into a duct system than centrifugal fans?
10 Give three reasons why axial flow fans have become increasingly used in recent years.

C Use each of the following words in a sentence:

1. rotate 2. continuously 3. curved 4. hub 5. outlet

Motor Cars

1. How the Engine Works

An engine that converts heat energy into mechanical work is called a heat engine, and the car engine is one type of heat engine. It derives heat from the burning, or 'combustion', of a fuel and converts this heat into useful work for driving the car.

The fuel used in the **vast majority** of car engines is petrol, which is one of the many products obtained from crude oil found in the earth. Petrol, when mixed with the right amount of air, will burn when a flame or spark is applied to it.

In the car engine, air mixed with petrol is taken into **a confined space** and **compressed.** The mixture is then ignited and it burns. In burning it heats the air, which expands, and the force of expansion is then converted into **a rotary movement** to drive the wheels of the car.

To be able to use this energy effectively we have to control the burning or combustion process and the force of expansion. Firstly, we need a tube, or 'cylinder', closed at one end, in which to compress and burn the petrol and air mixture. Then we need a piston which can slide freely in the cylinder, and which can be driven outwards by the force of expansion. To convert the outward movement of the piston into a rotary movement we must join it by a connecting rod to a crankshaft. We need one passage for the entry of the mixture into the cylinder and another to let out the used gases. To control the entry of the mixture and the exhaust of the gases we need **valves,** and these are called the **inlet** and **exhaust** valves. Finally, we need some means of igniting the mixture in the top of the cylinder, the part called the combustion chamber, and for this we use a sparking plug.

By timing the opening and closing of the valves and by timing the arrival of the spark we can control **the whole sequence of**

events and make the piston move in and out over and over again.

THE CYLINDERS

Motor-car engines may have four, six, or eight cylinders. Look at figure 18. These cylinders are usually **mounted** in a cylinder

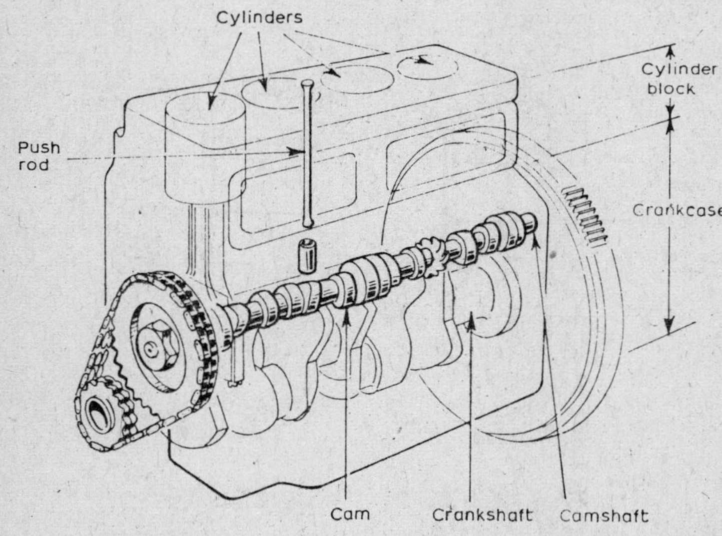

Figure 18

block on top of the engine. Beneath the cylinder block is the crankcase, which contains two shafts, the crankshaft and the camshaft. As you have read, the crankshaft is revolved by the outward movement of the pistons in the cylinders. This rotary movement of the crankshaft transmits the power developed by the engine through the gearbox to the driving wheels and sets the car in motion.

When the crankshaft rotates it also causes the **rotation** of the camshaft, which lies alongside it in the crankcase. As the camshaft rotates, it pushes up rods alongside each cylinder to open and shut the valves at the top of the cylinder.

There are two valves to each cylinder. The inlet valve lets air and petrol into the combustion chamber of the cylinder when it is opened. When the exhaust valve is opened the gases formed after the combustion in the chamber are allowed to escape. These gases (carbon monoxide) are led away from the car through an exhaust pipe.

Let us examine the action of one of the cylinders in more detail. Look at figure 19. In (*a*) the piston is near the top of the

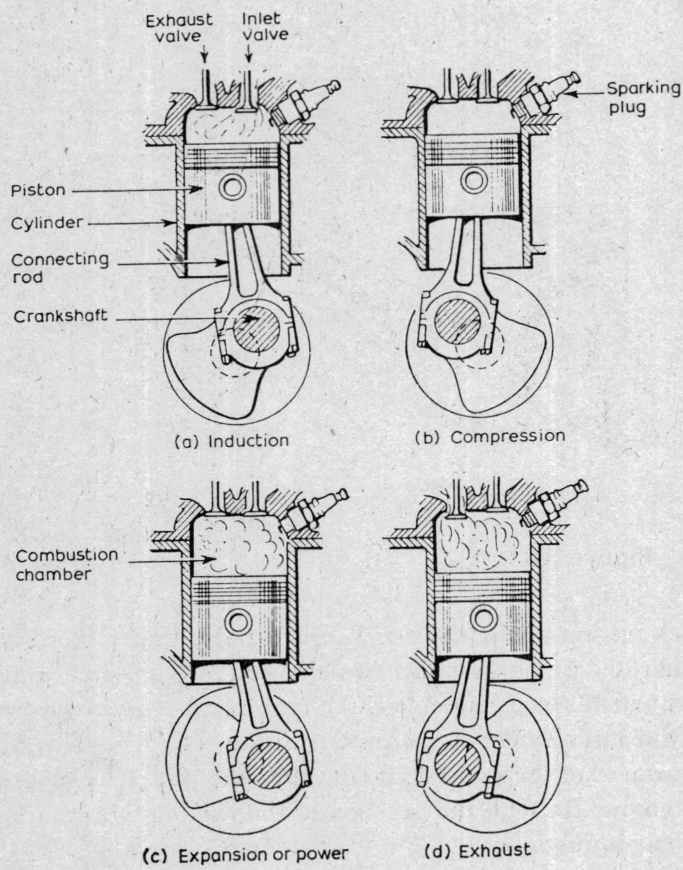

Figure 19

cylinder with the inlet valve open and the exhaust valve closed. If the crankshaft is turned, either by hand or by the starter motor, the piston is drawn down by the connecting rod and **a**

charge of petrol–air mixture rushes in. When the piston reaches the bottom of **its stroke** the inlet valve is closed by the action of a spring. This stroke is called the 'induction' stroke.

In (*b*) both valves are closed and the crankshaft forces the piston up to compress the mixture in the top of the cylinder. This, then, is called the 'compression' stroke. Towards the end of the compression stroke a spark from the sparking plug causes the mixture to ignite.

In (*c*) we see that the heat of combustion has caused a rapid rise in pressure in the combustion chamber and this has forced the piston down. Through the connecting rod the piston causes the crankshaft to continue to rotate. This stroke is called the 'expansion' or 'power' stroke, and we can say now that the engine has 'fired'. At the end of this stroke, as the crankshaft rotates, causing the camshaft alongside it to continue to rotate, one of the cams on the camshaft pushes up the rod, which causes the exhaust valve to open, allowing the exhaust gases to escape.

In (*d*) we see the fourth and final stroke, the 'exhaust' stroke. The exhaust valve has been forced open by the rotation of the camshaft, and the crankshaft, continuing to rotate, drives the piston back up the cylinder, forcing out the exhaust gases. At the end of this stroke the exhaust valve is closed by the action of a spring, and the camshaft, continuing to rotate, pushes up a second rod to force the inlet valve open. Now the cylinder will receive another charge of petrol–air mixture, and the sequence of four strokes, 'induction', 'compression', 'expansion' or 'power', and 'exhaust', will start again.

Because there is a four-stroke sequence, or **'cycle',** in this type of internal combustion engine it is called a four-stroke engine. There are also two-stroke engines used, for example, for motor scooters, and for some motor cycles.

Let us think of a car with four cylinders. Remember that it is only on the expansion stroke that power is transmitted to make the crankshaft rotate. Let us number the cylinders: 1, 2, 3, 4. They may transmit power in this order: 1, 2, 4, 3. This means that when number 1 cylinder is on the expansion stroke, number 2 is on the compression stroke, number 4 is on the induction stroke, and number 3 cylinder is on the exhaust stroke. The four pistons, moving up and down inside their cylinders in this order, push down rods connected to the crankshaft at different times and at different points along its shaft. This keeps the crankshaft revolving and the crankshaft, **when the gears are engaged,** keeps the car moving.

Adapted from *The Shell Book of How Cars Work* by Gordon Walmsley (John Baker and Shell-Mex & B.P. Ltd.)

Explanation of words in bold type

vast majority greatest number.
a confined space a space kept to a small size.
compressed pressed together.
a rotary movement a movement going round and round.
valves instruments which let gases or liquids pass in one direction only.
inlet, exhaust see Ventilation 3.
by timing by arranging the time of.
the whole sequence of events the order in which things happen.
mounted in fitted into.
rotation the motion of going round and round, or 'revolving'.
a charge see Metals 2.
its stroke its movement up and down.
cycle a sequence of events which happens again and again.
when the gears are engaged when the gears are connected to the driving wheels of the car.

Exercises

A Use these words to complete the following sentences:
air, camshaft, charge, converted, crankcase, gases, ignited, petrol, pistons, pressure, two-stroke, valve.

1 In a car engine heat is . . . . into useful work.
2 The fuel used in a car engine is . . . . and . . . .
3 The . . . . slide up and down inside the cylinders.
4 The mixture in the combustion chamber is . . . . by a sparking plug.

5 There are two shafts inside the . . . .
6 The valves at the top of the cylinders are opened and shut by the action of the . . . .
7 During the induction stroke a . . . . of petrol–air mixture enters the combustion chamber.
8 The piston is forced down by a rapid rise in . . . . during the expansion stroke.
9 When the exhaust . . . . is forced open the . . . . caused by combustion are allowed to escape.
10 Motor scooters use a . . . . engine.

B Answer these questions.

1 What three things are required to make the combustion in the combustion chamber?
2 What are the two valves at the top of each cylinder called?
3 How is the downward movement of the pistons converted into a rotary movement?
4 What does the crankcase contain?
5 What makes the camshaft revolve?
6 What function of the camshaft is mentioned in the passage?
7 What happens in a cylinder during the induction stroke?
8 What forces the piston down the cylinder?
9 When do we say that the engine has fired?
10 What stroke follows the induction stroke in a four-stroke engine?
11 How is the power developed by the engine transmitted to the driving wheels of the car?
12 What happens when the exhaust valve in the cylinder is forced open?

2. The Fuel and Ignition Systems

The fuel system includes a fuel tank, a fuel pump, and a carburettor.

The fuel pump may be operated either mechanically from the engine camshaft or electrically. Its **function** is to pump petrol from the petrol tank and deliver it to the carburettor. It contains a **fine filter** to **exclude** particles of dust or dirt which may have found their way into the tank.

The carburettor is **mounted on** the inlet pipe, or inlet 'manifold', which leads to the inlet valves of each cylinder. Its function is to 'carburet', or mix, the required amounts of petrol and air for combustion in the cylinders. It **regulates** automatically the proportions of petrol and air and also allows the driver to control the rate of delivery of the mixture, and so vary the speed of the engine.

See figure 20. This is a simple type of carburettor. It makes use of a fuel reservoir or 'float chamber' to which petrol is

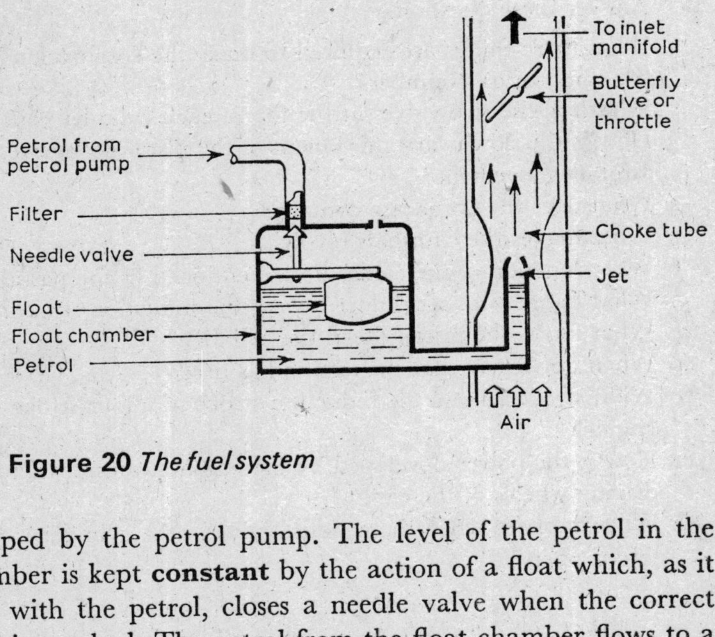

Figure 20 *The fuel system*

pumped by the petrol pump. The level of the petrol in the chamber is kept **constant** by the action of a float which, as it rises with the petrol, closes a needle valve when the correct level is reached. The petrol from the float chamber flows to a small jet situated in a narrow tube called the choke tube. When the engine is running, i.e. when the pistons are moving up and down in the cylinders, air is drawn in and passes through the choke tube. Here it mixes with the petrol and passes out into the inlet manifold of the engine. The amount of mixture allowed to pass is controlled by a butterfly valve, or 'throttle', situated in the carburettor outlet. This valve is operated by the **accelerator pedal.** An air cleaner may be connected to the air inlet to remove dust from the air and prevent it getting into the cylinders and down into the engine lubricating oil, where it can cause increased engine wear.

The proportion of air and petrol required at varying engine speeds is controlled automatically. When starting a cold engine a much richer mixture, one with a higher proportion of petrol to air, is required. This is obtained by **restricting** the amount of air entering the carburettor. In most cars the driver does this by pulling out a knob called the choke and this partly closes the tube carrying air to the cylinders. The driver must remember to push the choke in again as soon as possible after starting to avoid damage to the engine and waste of fuel.

THE IGNITION SYSTEM

The function of the ignition system is to provide the spark in the combustion chamber to ignite the mixture of petrol and air

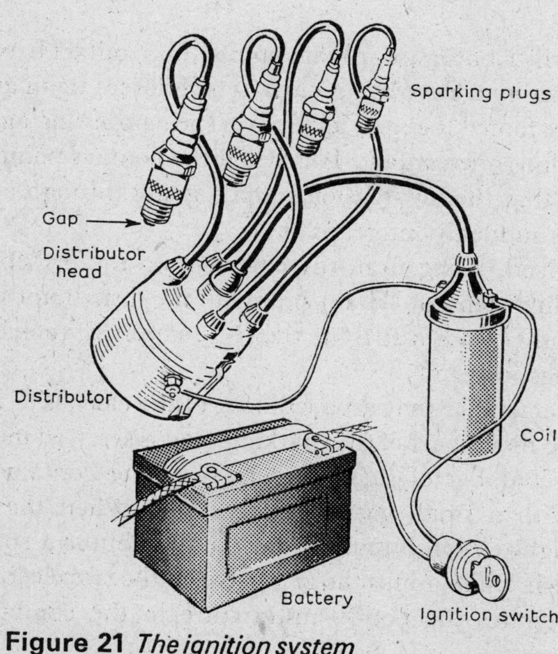

Figure 21 *The ignition system*

at the right instant. The system normally **comprises** a coil, a distributor, and sparking plugs.

The coil produces the high-voltage impulse required to make the spark at the sparking plugs. It really consists of two coils of insulated wire wound one around the other so that the number of turns in the inner, 'secondary', coil is much greater than that in the outer, 'primary', coil.

A low voltage is supplied to the primary coil and suddenly interrupted. At that moment an **impulse** at very much higher voltage is induced in the secondary coil. Voltages of 6 or 12 volts supplied to the primary, low-tension, coil result in voltages of 10,000 volts or more being obtained from the secondary, high-tension, coil. When the car is started the low-tension supply is provided by the car's battery.

Remember that the low-tension supply must be broken suddenly to produce the high-tension supply from the secondary coil. This is done in an instrument called the distributor. The distributor is placed between the coil and the sparking plugs.

Inside the distributor are contact points. The contact between them is broken by **a revolving cam,** which forces them apart. Then they are forced together again by a spring pressing against the arm holding one contact. It is when the contact points are forced apart that the low-tension supply, going through a lead to the coil, is suddenly interrupted.

A second lead, going from the coil back to the distributor, conveys the high-tension (H.T.) supply to the distributor head. From there other leads distribute the H.T. supply to each of the sparking plugs in turn.

You will remember that each cylinder of the car has a sparking plug. The metal part of the sparking plug is screwed into the combustion chamber of the cylinder. The plug contains two electrodes with a small gap between them. When the plug receives a high-tension impulse from the distributor a spark is caused as the impulse jumps the gap between the two electrodes. This spark ignites the petrol–air mixture in the combustion chamber.

Adapted from *The Shell Book of How Cars Work* by Gordon Walmsley (John Baker and Shell-Mex & B.P. Ltd.)

Explanation of words in bold type

function	purpose, or duty. ('duty' is used for persons: The soldier's duty is to defend his camp. 'function' is used for things.)
a fine filter	1. a filter allows liquids to seep through it, but holds back solids. 2. 'fine' here means narrow. The spaces through which the petrol may pass are very narrow so that even very small particles of dirt, such as pieces of dust, are held back.
exclude	keep out. Opposite – include.
mounted on	fixed on to.
regulates	controls.
constant	see Ventilation 1.
accelerator	see Ventilation 1.
pedal	a control operated by the foot.
restricting	limiting.
comprises	is composed of.
impulse	a flow of electric current lasting only for a moment.
a revolving cam	a cam that is turning around on its fixed central point.

Exercises

A Use these words to complete the following sentences: accelerator, carburettor, contact, distributor, excluded, float, gap, head, manifold, secondary, smaller.

1 Petrol is conveyed to the .... by the fuel pump.
2 Dust is .... from the petrol by a filter.
3 Petrol–air mixture is led to the inlet valves of the cylinders through an inlet ....
4 A .... is used to keep the level of the petrol in the carburettor constant.
5 The driver can operate the throttle in the carburettor outlet by pressing the ....
6 The primary coil consists of a much .... number of turns than the secondary coil.
7 The high-tension supply comes from the .... coil.
8 The low-tension supply is interrupted when the .... points in the .... are forced apart.

9 High-tension supply to the sparking plugs comes through leads from the distributor . . . .
10 Between the two electrodes of the sparking plug there is a . . . .

B Answer these questions.

1 How is dust kept out of the petrol which is led into the carburettor?
2 How can the driver vary the speed of the engine?
3 What is the function of the needle valve in the float chamber?
4 Why must dust be prevented from getting into the lubricating oil in the engine?
5 How may a richer mixture of fuel supply be obtained for starting the car on a cold morning?
6 What is the function of the coil?
7 How is a very high voltage induced in the high-tension coil?
8 How is the contact broken inside the distributor?
9 Why, do you think, the distributor was given its name?
10 How is the spark produced from the sparking plug?

C Use each of the following words in a sentence:

1. compress 2. confine 3. pressure 4. expands 5. revolve

3. Cooling and Lubricating the Engine

The combustion of the mixture in the engine naturally makes a great deal of heat and the engine must not be allowed to become overheated. The function of the cooling system is to control this heat.

The engine may be either water or air cooled, but the vast majority are water cooled. Water in the water jacket surrounding the cylinders picks up the surplus heat from the engine and the heated water rises. It then **circulates** through the radiator, where it is cooled by the passage of air past the radiator tubes, and returns to the bottom of the water jacket. A fan, which is driven from the engine crankshaft, assists cooling by increasing the draught through the radiator at low speeds. A pump may be used to help in circulating the water.

The engine must also be prevented from running too cool and therefore the temperature of the water is normally controlled within certain limits by the action of a **device** known as

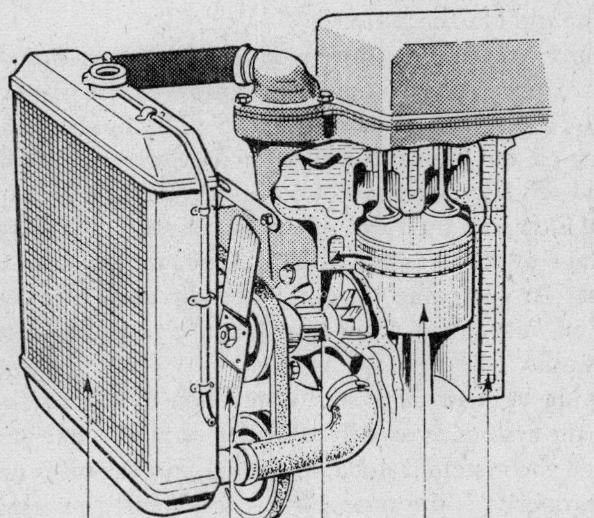

Figure 22 *The cooling system*

a thermostat. This has the effect of varying the circulation of the water so that, for instance, when the engine is first started less water is allowed to circulate and the engine warms up quickly.

The cooling system can, of course, freeze up in very cold weather, and this can cause considerable damage. For this reason it is advisable to add a good anti-freeze **solution,** one that can be left in throughout the summer, which will prevent freezing in winter and protect the system from corrosion all the year round.

THE LUBRICATION SYSTEM

Lubrication is necessary for two main reasons: to reduce friction and wear in the engine, and to help carry away waste heat from

the bearings, in which the crankshaft and camshaft revolve inside the crankcase, from the pistons, and from the other **working parts.** In addition, it helps the piston rings to seal in the combustion gases, to prevent them escaping down into the cylinders. (These piston rings are rings fitted into grooves around the top of each piston.)

The majority of engines have what is known as a 'force feed', or pressure, lubricating system. The bottom of the crankcase is enclosed by a sheet-metal pan, or 'sump'. This holds the engine lubricating oil and has a drain plug through which used oil may be drained off. A pump draws oil from the sump and pumps it through a filter and then through channels, called 'oilways', to the bearings of the crankshaft and camshaft, and also to the valve gear. At the same time, oil flowing from the bearings forms an oil mist in the crankcase and this mist lubricates the cylinder walls.

A warning light or oil gauge, controlled by the pressure of the oil in the system, is usually provided so that the driver will know when the system is not functioning and the oil is not circulating properly. If the system is not functioning properly the working parts may **seize up** and cause serious damage to the car engine. It is, therefore, essential for the driver to stop the engine as soon as he sees his warning light come on.

Adapted from *The Shell Book of How Cars Work* by Gordon Walmsley (John Baker and Shell-Mex & B.P. Ltd.)

Explanation of words in bold type

circulates	goes around.
device	instrument planned for a special purpose.
a solution	usually means a liquid in which a solid has been dissolved.
bearings	support a shaft in its correct position.
working parts	parts that move when the engine is running.
seize up	lock together.

Exercises

A Use these words to complete the following sentences:
anti-freeze, circulate, crankshaft, friction, mist, rises, seize up, sump, thermostat, water.

1 There are two systems used to cool car engines: air cooling and .... cooling.
2 When the water around the cylinders becomes heated it ....
3 The .... drives a fan which helps to cool the water when the engine is running slowly.
4 The .... prevents the engine from running too cool.
5 .... solution should be added to the water in the radiator during very cold weather.
6 The lubrication system reduces .... in the engine.
7 The engine oil may be drained off through a plug in the bottom of the ....
8 The cylinder walls are lubricated by an oil .... coming from the crankcase.
9 If oil is not circulating through the engine when it is running the engine will ....
10 If the oil does not .... properly the driver is informed by a warning light.

B Answer these questions.

1 What is the function of a car's cooling system?
2 Give *two* ways in which the water is cooled as it passes through the radiator.
3 Why is a smaller amount of water circulated around the cylinders when the engine is first started?
4 Why is anti-freeze solution sometimes added to the water in the radiator?
5 From which parts of the engine does the lubrication system help to carry away heat?
6 What is the function of the piston rings?
7 How can the engine oil be drained off?
8 Why is it important for the driver to take notice of the oil warning light?

C Use each of the following words in a sentence:

1. lubricate 2. filter 3. regulate 4. accelerator 5. friction

4. The Brakes

The brakes function by absorbing in friction the energy possessed by the moving car. In so doing they convert the energy into heat.

There are two types of brakes, the drum brake and the disc brake. Either or both types may be fitted, but where both types are used it is usual for the disc brakes to be fitted to the front wheels.

DRUM BRAKES

The drum brake consists of a pair of semicircular brake shoes mounted on a fixed back plate and situated inside a drum. This drum is fixed to the road wheel and rotates with it. One end of each shoe is on a **pivot** and a spring holds the other end in contact with the piston of a **hydraulic** cylinder. (In front brakes it is usual to use two hydraulic cylinders in order to **equalise** the pressures **exerted** by the shoes. See figure 23.) Each shoe is

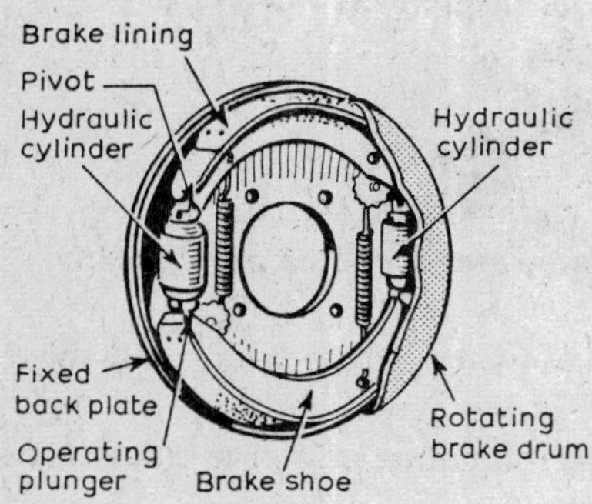

Figure 23 *Drum brake*

faced with material, known as brake lining, which produces high frictional resistance.

The hydraulic system comprises a master cylinder and the slave cylinders, which are the cylinders on the road wheels. The slave cylinders are connected to the master cylinder by tubing and the whole system is filled with hydraulic fluid. A piston in the master cylinder is connected to the brake pedal, so that when the driver **depresses** the pedal the fluid is forced out to each slave cylinder and operates their pistons. The fluid pushes the pistons out of their cylinders. They, in turn, push against the inner ends of the brake shoes and force them against the brake drums in each wheel. We say that the brakes are on. This friction of the shoes against the drums, which are fixed to the road wheels, slows down or stops the car. As the brake pedal is allowed to come up, the hydraulic fluid returns to its original position, the pistons **retract,** and a spring attached to each brake shoe returns it also to its original position, **free of** the brake drum. Now we say that the brakes are off.

The brakes may also be operated by mechanical **linkages** from the foot pedal and handbrake lever. **Common practice** is to operate both front and rear brakes hydraulically with a secondary mechanical system operating the rear brakes only from the hand lever. One of the great advantages of hydraulic operation is that the system is self-balancing, which means that the same pressure is automatically produced at all four brakes, whereas mechanical linkages have to be very carefully adjusted for balance. Of course, if more pressure is put on one of the brakes than on the others there is a danger that the car will **skid.**

The mechanical linkage operating on the rear brakes is a system of rods or cables connecting the handbrake lever to the brake-shoe mechanisms, which work entirely **independently of** the hydraulic system.

Drum brakes are **prone** to a reduction in the braking effort, known as 'fade', caused by the overheating of the linings and the drum. Fade can affect all or only some of the brakes at a time, but it is not permanent, and full efficiency returns as soon as the brakes have cooled down. However, fading is unlikely to occur except after the brakes have been used repeatedly in slowing the car from a high speed or after braking continuously down a steep hill. **Descending** such a hill, it would have been preferable to use engine braking by changing down into a **lower gear.** Drum brakes can be made less prone to fade by improving the cooling arrangements, by arranging for more air to be deflected over them, for example.

DISC BRAKES

The disc brake consists of a steel disc with friction pads operated by slave hydraulic cylinders. The steel disc is attached to the road wheel and rotates with it. Part of this steel disc is enclosed in a caliper. (See figure 24.) This caliper contains two friction

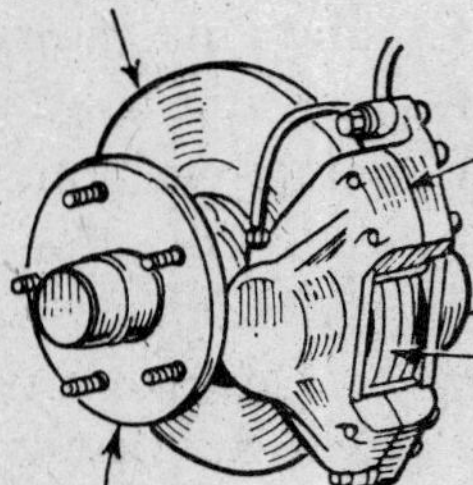

Figure 24 *Disc brake*

pads, one on each side of the disc, and two hydraulic cylinders, one outside each pad. The pads are normally held apart by a spring, but when the driver depresses the brake pedal, pistons from the hydraulic cylinders force the pads against the sides of the disc. Because the disc is not enclosed all the way round, the heat generated when the brakes are **applied** is **dissipated** very much more quickly than it is from brake shoes which are entirely enclosed inside a drum. This means that disc brakes are less prone to fade than drum brakes.

Adapted from *The Shell Book of How Cars Work* by Gordon Walmsley (John Baker and Shell-Mex & B.P. Ltd.)

Explanation of words in bold type

pivot — the point at which one end of the brake shoe is fixed. The brake shoe can swing from this fixed point.

hydraulics	is the science of the flow of fluids.
equalise	make equal.
exerted	put on.
depresses	presses down.
retract	go back.
free of	not touching.
linkages	anything that **links** parts together, or **connects** them. Here the linkages are rods which pull on one another in turn when the driver pulls on the handbrake or depresses the footbrake.
common practice	the usual system.
skid	slide over the surface when the tyres are not gripping the road.
independently of	separately from.
prone	He is prone to get colds in winter. He is **likely** to get colds in winter. OR: It is **probable** that he will get colds in winter.
descending	going down (cf. **de**pressing).
lower gear	most cars have three or four gears, 1st, 2nd, 3rd, 4th, for driving forward, and one for reversing, simply called 'reverse'. On a four forward-gear car, 4th gear is called top gear and is used when the car is running at normal or fast speeds. The other gears are the lower gears.
applied	put on.
dissipated	see Ventilation 1.

Exercises

A Use these words to complete the following sentences: cooling, fade, hydraulic fluid, mechanically, more, on, pads, pressure, semicircular, tubing.

1 Brake shoes are . . . . in shape.
2 . . . . connects the master cylinder to the slave cylinders.
3 The pistons in the cylinders are pushed out by . . . . when the brake pedal is depressed.
4 The brakes are . . . . when the brake shoes are forced against the drum.

5 When the driver applies the handbrake the rear brakes are operated . . . .
6 To avoid skidding it is important to apply the same braking . . . . on all four wheels.
7 . . . . is when the efficiency of the brakes is reduced because the linings and drum are overheated.
8 . . . . can make brakes less prone to fade.
9 In the disc braking system . . . . press on each side of the disc when the brakes are applied.
10 Drum brakes are . . . . prone to fade than disc brakes.

B Answer these questions.

1 What are the two types of brakes used in cars?
2 What is 'brake lining'?
3 What is the purpose of the hydraulic fluid in the braking system?
4 Which brakes are usually operated by the handbrake lever?
5 What advantage have hydraulically operated brakes over mechanically operated brakes?
6 Why is it important to apply pressure equally to each wheel when braking?
7 Give *two* occasions when fade is likely to occur.
8 How can brakes be made less prone to fade?
9 What is contained inside the caliper in the disc braking system?
10 Why are disc brakes less prone to fade than drum brakes?

C Use each of the following words in a sentence:

1. circulate 2. working parts 3. piston 4. rear 5. vary

5. Electric Cars

Since the 1890s the vast majority of mechanically operated road vehicles have used internal-combustion engines. By comparison, steam engines used to **propel** road vehicles were found to be bulky, noisy, dirty, and more difficult to refuel, petrol and diesel oil being more conveniently stored and supplied to a vehicle than solid fuel. However, road vehicles operated entirely by electricity have also been used for many years alongside cars with internal-combustion engines. These electric vehicles are of two types: trolley buses, which receive their electricity supply by being connected by rods to overhead electric cables, and small **delivery vans,** which are electrically charged by batteries. Now attempts are being made to increase the proportion of electrically powered vehicles on the roads, particularly by manufacturing an ordinary motor car **thus powered.**

Electric cars would have two great advantages over today's motor cars. They would be quieter and cleaner. The second advantage is of particular importance since **pollution of the atmosphere** in cities by the exhaust carbon monoxide gases from cars is a **threat** to health.

There have been two main **impediments** to the **mass production** of electric cars: the heavy weight of the batteries required and the limited range of these batteries before they need recharging. Trolley buses and electric delivery vans operate only within cities. The trolley buses, although capable of reasonable speeds, can only follow routes where overhead cables have been **erected,** and obviously not enough of these cables could be supplied to satisfy all motor-car travellers. The delivery vans cannot travel fast, nor can they travel far without their batteries needing a recharge, a slower operation than simply refilling a petrol tank at a garage.

Recently, however, research workers have produced in laboratories new types of batteries which could overcome the two disadvantages mentioned. At present, batteries fitted into motor cars are of the lead–acid type. They do a very limited amount of work, but are heavy and bulky. One of the new types of batteries has current generated by air reacting with zinc to form zinc oxide. Air is pumped under pressure and the working fluid circulated past zinc plates. These zinc–air batteries can give five times more power for their weight than lead–acid batteries.

Research workers of the General Motors company are working to develop a similar battery using hydrogen and oxygen reacting with zinc. **Prototype** electric cars produced by General Motors will have a speed of up to **120 m.p.h.** and a range of up to 80 miles before the batteries will need recharging.

Ford company researchers are working on an even more efficient battery. It will be made mainly of sodium, sulphur, and aluminium, which are all reasonably cheap and plentiful materials. It may give more than twice the power for weight than the zinc–air battery, i.e. more than ten times that of the lead–acid battery. The range of the sodium–sulphur battery may be over 250 miles driving in towns and possibly up to 500 miles driving outside towns.

How will these new types of batteries be recharged? There are two possibilities. One is that each driver will connect his batteries to electric supply points at home to recharge them every night after he has used his car. Alternatively, or in addition to this, he can simply exchange his **spent batteries** for recharged batteries at a garage.

If, in fact, all cars are going to be operated entirely electrically the demand for electricity will increase tremendously. It has been calculated that in the United States cars **consume** energy **equivalent** to the country's entire electricity generating capacity.

In an electric car possibly one or more batteries will be situated near each **axle** and will operate driving motors actually incorporated into the axle. Therefore, no **transmission unit** will be required. In fact, electric cars will be much simpler than present cars. There will also be no need for the present engine, gearbox, fuel tank, exhaust unit, radiator, or cooling system, and as the engine will be much quieter than the petrol-driven engine, the problem of **insulating** the car against noise will be much reduced.

Explanation of words in bold type

propel set in motion.
delivery vans vans that deliver goods, such as bread and milk, to people's houses.
thus powered powered in this way (i.e. electrically powered).
pollution of the atmosphere making the air dirty.
threat danger.

impediments obstacles. Factors to stop progress.
mass production manufacture of very large numbers.
erected put up.
prototype cars are the first working models of a new type of car. These models are made for testing and demonstration.
120 m.p.h. 120 miles per hour.
spent batteries have been used until they no longer function.
consume use up.
equivalent equal.
axle the shaft that connects the two front or two rear road wheels.
transmission unit the parts that pass on, or 'transmit', the engine power to the driving wheels of the car. In most cars today engine power is developed in the front and transmitted the length of the car to the rear wheels.
insulating see Ventilation 2.

Exercises

A Use these words to complete the following sentences: generated, insulation, lead–acid, limited, polluted, prototype, recharging, solid, spent, ten.

1 Steam engines use . . . . fuel.
2 The atmosphere is . . . . by exhaust gases from today's cars.
3 Until recently electric road vehicles have had only a . . . . range.
4 The batteries in delivery vans need . . . . frequently.
5 At present motor cars have . . . . batteries fitted into them
6 A battery made mainly of sodium, sulphur, and aluminium may give more than . . . . times the power for weight than a lead–acid battery.
7 The General Motors company are producing . . . . electric cars with speeds of up to 120 m.p.h.
8 A driver will be able to get recharged batteries in exchange for . . . . batteries at a garage.
9 Because electric cars will be less noisy than today's cars, ... problems will be reduced.

10 A great increase in the amount of electricity . . . . will be necessary if electric cars are mass produced.

B Answer these questions.

1 Why are very few road vehicles driven by steam?
2 How are trolley buses powered?
3 Why will electric cars be cleaner than cars driven by internal-combustion engines?
4 Give *two* reasons why electric cars have not yet been mass produced.
5 What will be the range of the prototype electric cars produced by the General Motors company?
6 What materials have Ford research workers used in their battery?
7 Why cannot all cars be operated like trolley buses?
8 Give *two* ways in which drivers may ensure that the batteries of their electric cars are always charged.
9 Why will there probably be no need for a transmission unit in an electric car?
10 Name *three* other parts of today's cars which will not be needed in an electric car.

C Use each of the following words in a sentence:

1. transmit   2. skid   3. prone   4. descend   5. generated

Welding

1. Oxy-acetylene Welding

The art of welding is probably as old as any of the processes for the working of metals, and the beautiful creations of the **old-time** blacksmith included many perfect examples of hand welding carried out at the **forge.** Fire welding, in which the surfaces to be joined are heated to a **pasty state** and then hammered together, is still practised to the limited extent necessary during the hand forging of small numbers of articles, but in all other cases **fusion** or electric methods are employed.

When parts are joined by fusion the surfaces being welded are heated until they melt and run together. If **a filler** is added this is melted along with the metals being joined and runs into the space which it is necessary to fill up. Since the melting temperature of mild steel is about 1,200° C., very high **local temperatures** are necessary to melt the metal at the point where the weld is to take place. The two chief methods of obtaining such temperatures are the oxy-acetylene flame and the electric arc.

Oxygen and acetylene, when mixed in suitable proportions, give a flame which in its hottest region **attains** a temperature of approximately 3,200° C. There are two methods of oxy-acetylene working known as the high-pressure and the low-pressure systems, these terms referring to the form in which the acetylene is supplied. In the high-pressure system both oxygen and acetylene are used from high-pressure cylinders. Oxygen cylinders are charged to a pressure of 120 atmospheres (approx. 1,800 lb. per sq. in.), but acetylene, owing to its **explosive tendency** when compressed, is supplied in cylinders in the form known as 'dissolved acetylene'. The cylinders are filled with a **porous material** in the form of a charcoal cement and charged with acetone. This liquid has the property of absorbing

25 times its own volume of acetylene for each atmosphere of pressure applied. The pressure of the acetylene is usually about 225 lb. per sq. in. and cylinder sizes vary from 50 to 250 cu. ft.

In the low-pressure system the acetylene is generated in a low-pressure generator by the action of water on calcium carbide, and is supplied to the welder's blowpipe at low pressure from a gas-holder **incorporated in** the generator. When this method is used the acetylene has to be cleaned by passing through a purifier, and to prevent the possibility of an explosion by oxygen or air blowing back and entering the generating plant, a back-pressure valve must be placed between the blowpipe and the gas-holder.

Before the high-pressure gas from the cylinders may be used in the blowpipe its pressure must be reduced, and the supply from the cylinder to the blowpipe is taken through a reducing valve which screws into the cylinder outlet.

Tubes from the oxygen and acetylene supplies lead to the blowpipe, the supply of each gas being controlled by a regulating valve. On the high-pressure system the gases pass to a mixing chamber and from there through the head of the blowpipe to the nozzle (or 'tip'). The function of the mixing chamber is to ensure that the gases are mixed in correct proportions (approximately equal) before being burnt at the tip.

In the low-pressure blowpipe the oxygen is led through an injector nozzle inside the body of the pipe and **in its passage** draws the low-pressure acetylene into the mixing chamber, giving it the necessary **velocity** to preserve a steady flame.

In order to vary the size of flame (and heat supply) necessary to weld varying thicknesses of metal, a selection of tips is available for the blowpipe. Each of these is marked showing **its gas consumption** in litres per hour, and a table supplied with the blowpipe shows which tip should be used to weld any required thickness of metal. The delivery pressure from the regulator must be varied according to the size of tip used, and instructions are supplied for obtaining the correct conditions.

The correct adjustment of the flame is important for reliable welds. If there is **an excess of** oxygen (oxidising flame) the weld will be oxidised. This will result in a large grain size, which means that where the metal is joined it will be weaker and more brittle than the rest of the metal. A carbonising flame is one in which there is an excess of acetylene (C_2H_2). When welding steel this will **tend** to give the steel in the weld a higher carbon content than **the parent metal,** resulting in a hard and brittle weld. For most welding operations a neutral flame must be used, and this is most easily attained by commencing with the carbonising flame as shown in figure 25a, and gradually reducing the acetylene supply until the large cone just fades

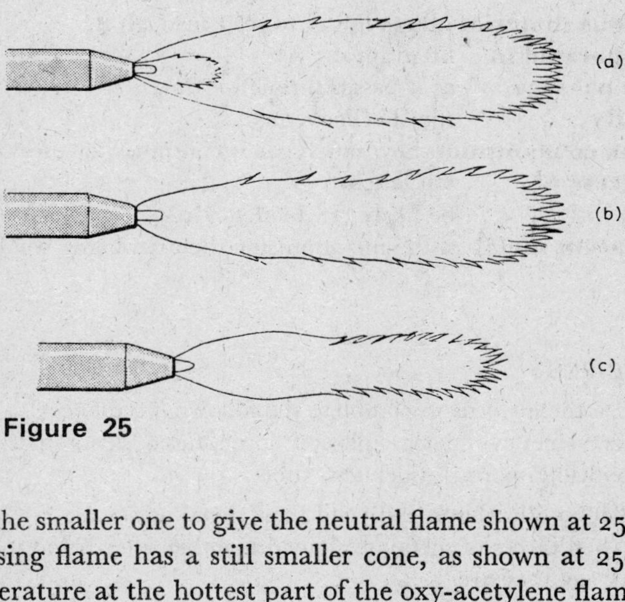

Figure 25

into the smaller one to give the neutral flame shown at 25b. The oxidising flame has a still smaller cone, as shown at 25c. The temperature at the hottest part of the oxy-acetylene flame (just outside the cone) is approximately 3,200° C.

Adapted from *Workshop Technology*, Part 3,
by W. A. J. Chapman (Edward Arnold)

Explanation of words in bold type

old-time blacksmith a blacksmith who worked a long time ago.

forge a workshop where the blacksmith worked, hammering heated metal.

a pasty state is when the metal is in the condition like that of paste (or glue), neither liquid nor solid.
fusion is when the weld is made only by melting the metal. No mechanical pressure is used to join the metal together.
a filler is a metal which fills any spaces there may be between the parts that are welded together.
local temperatures are temperatures reached at a particular part of the metal.
attains reaches.
explosive tendency something that has a **tendency** to explode is **likely** to explode, or **prone** to explode.
a porous material allows liquid to seep through it.
incorporated in fitted into.
in its passage as it passes through.
velocity speed of movement.
its gas consumption how much gas it consumes, or uses up.
an excess of too much.
tend be likely (cf. tendency).
the parent metal is the metal outside the area being welded.

Exercises

A Use these words to complete the following sentences: acetylene, cylinders, explosion, forge, fused, generated, oxidising, porous, thickness, tubes.

1 Blacksmiths shape and weld their metal at a . . . .
2 When pieces of metal are joined together after being melted we say they are . . . .
3 In the high-pressure method of oxy-acetylene welding, acetylene is supplied in . . . .
4 The charcoal cement in the acetylene cylinders is . . . .
5 Acetylene can be . . . . by the action of water on calcium carbide.
6 A back-pressure valve prevents air blowing back into the acetylene generator and causing an . . . .
7 Oxygen and acetylene are supplied to the blowpipe through . . . .
8 The tip of the blowpipe may be varied according to the . . . . of the metal to be welded.

9 An . . . . flame will result in a weakness of the metal in the weld.

10 An excess of . . . . in the flame will result in a carbonising flame.

B Answer these questions.

1 What is the difference between fire welding and fusion welding?

2 Why are very high temperatures necessary to weld mild steel?

3 To what pressure is the oxygen charged for oxy-acetylene welding?

4 What is 'dissolved acetylene'?

5 Why is acetylene supplied in the form of dissolved acetylene in the high-pressure system of oxy-acetylene welding?

6 What is the function of the reducing valve mentioned in the passage?

7 What happens inside the mixing chamber?

8 What does a low-pressure blowpipe contain that a high-pressure blowpipe does not?

9 What is marked on each tip that can be fitted on to a blowpipe?

10 Give *two* reasons why a weld may be brittle.

11 Why is it sometimes necessary to change the tip on a blowpipe?

12 How can a welder attain a neutral flame?

2. Electric-arc Welding

In electric-arc welding molten metal from a welding rod is cast into the space between the metals which are being fused together. No blowpipe is used. The heat necessary to melt the metal and welding rod is obtained from an electric arc struck between the rod (or 'electrode') and the work, a temperature of 3,500°–4,000° C. being obtained near the **crater** of the arc.

Arc welding may be operated with direct, or alternating, electric current. In the direct-current (D.C.) welding sets the current is generated by a special generator which may be driven by an electric motor or by some form of engine. Electrically driven generators are usually employed in cases where the set can be permanently installed, and where a supply of current is available. Engine-driven sets are most commonly driven by a petrol or diesel engine and may be of the permanently fixed type or may be mounted on wheels so that they can be moved from place to place.

Alternating-current (A.C.) welding sets derive their current from an electrical **transformer** which takes in current at mains voltage and transforms it to the voltage necessary for the arc (80–100 volts). Transformer sets are cheaper and simpler than generator sets, and since there are no moving parts, **maintenance is almost nil.** These advantages are to some extent offset by the **superiority** of D.C. sets for welding cast iron and **non-ferrous metals** and the necessity of always using covered electrodes when working with alternating current.

The welding rods are, in fact, electrodes, and are made in various lengths from 8 in. to 18 in. and in diameters from $\frac{1}{8}$ in. upwards. **Their composition** varies according to the work for which they are required and they may be mild steel, cast iron, bronze, and other special compositions for particular work. Electrodes are nearly always coated with chemicals or fluxes which enable a steadier arc to be maintained, and, by fluxing and covering the weld, promote improved quality in the weld. Fluxing is the process of preventing the metal surfaces which are being joined from oxidising while they are being heated. A good weld cannot be made between oxidised surfaces. Therefore, while the surfaces are being melted, a flux made up of liquid **siliceous** matter covers the surfaces and prevents oxides from forming on them.

Electrodes may be lightly, medium, or heavily coated. The lightly coated rods have a thin layer of a chemical such as lime which helps to **stabilise** the arc but has little effect on fluxing or slag formation. Medium and heavily coated rods have a covering of paste containing siliceous matter which melts and fluxes the weld, at the same time depositing a heavy layer of slag over the finished weld, preventing oxidation and reducing the rate of cooling. The covering also contains substances which give off gases and **envelop** the molten crater of the weld, to prevent atmospheric oxidation while the operation is in progress. These rods assist and shield the arc because the covering melts at a slower rate than the metal core, with the result that the coating **projects** beyond the molten end of the electrode metal. (See figure 26.)

An electrode holder provides means for clamping and holding the rod in a suitable position and has the necessary electrical connection to the welding set. The **luminous intensity,** of

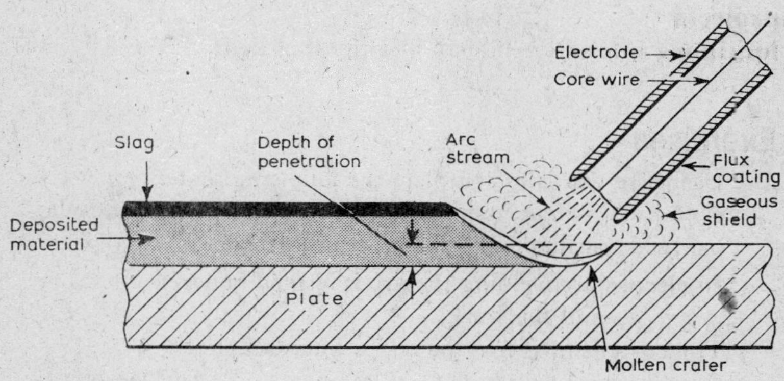

Figure 26

both the arc and the oxy-acetylene flame, makes suitable protection for the eyes necessary. With the arc an additional danger is the ultra-violet radiation which comes from it, necessitating a shield for the face and gloves on the hands. Goggles with green or blue glasses are usually worn for oxy-acetylene welding, while special shields are supplied for operating with the arc.

Adapted from *Workshop Technology*, Part 3
by W. A. J. Chapman (Edward Arnold)

Explanation of words in bold type

crater a pit (we also call the pit left in the ground by an earthquake a pit).

a transformer changes, or 'transforms', the voltage of an electrical supply.
maintenance is almost nil almost no maintenance is required.
superiority from the adjective 'superior' (better). Opposite: 'inferior'.
non-ferrous metals are metals not containing iron.
Their composition what they are made of.
siliceous contains silicon.
to stabilise to keep steady.
envelop form a cover around, as an envelope forms a cover around a letter.
projects sticks out.
luminous intensity intense brightness of light.

Exercises

A Use these words to complete the following sentences: chemical, current, diameters, direct, electrically, electrode, oxidised, permanently, shield, transformers.

1 In electric-arc welding the arc is formed between the .... and the work.
2 The current used may be either alternating or ....
3 Generators to supply D.C. may be .... driven or engine-driven.
4 The electrically driven generators are usually installed ....
5 .... supply current for A.C. welding sets.
6 A .... of between 80 and 100 volts is necessary for electric-arc welding.
7 The lengths and .... of electrodes vary.
8 If the surfaces to be joined are .... the weld will not be strong.
9 The arc can be stabilised by coating the electrode with a ....
10 A gaseous .... is often formed to prevent atmospheric oxidation during electric-arc welding.

B Answer these questions.

1 What is the arc which is formed in electric-arc welding?
2 What supplies the current in: (*a*) D.C. welding; (*b*) A.C. welding?

3 Why is very little maintenance required for transformer sets?
4 For which kind of welding are D.C. sets better than A.C. sets?
5 What is the purpose of siliceous matter in an electrode coating?
6 What *three* things may prevent oxidation in electric-arc welding?
7 Describe a heavily coated electrode.
8 Name *two* ways in which a welder may be injured when working.
9 What *two* things should a welder use to protect himself when making an electric-arc weld?
10 What equipment is used: (*a*) in oxy-acetylene welding that is not used in electric-arc welding; (*b*) in electric-arc welding that is not used in oxy-acetylene welding?

C Use each of the following words in a sentence.

1. fused 2. dissolved 3. layer 4. consumption 5. adjust

3. Welding Practice

Welding practice varies widely according to the type of metal to be welded and the kind of joint required. Let us consider the making of one kind of joint, a butt joint in steel, first using oxy-acetylene equipment.

A butt joint is used for welding two metal plates or bars in the manner shown in figure 27. The plates or bars to be joined may be of any thickness from $\frac{1}{32}$ **in. upwards.** In dealing with such work there is a choice between two methods of working: (*a*) the **leftward** (or forward), and (*b*) the **rightward** (or backward) method. In the leftward the weld is made working from right to left, while the reverse direction is followed for rightward working. In each case the filler rod, being held in the left hand, is carried to the left of the blowpipe. The leftward method is found most advantageous on steel plates up to $\frac{1}{4}$ in. thick and on non-ferrous metals, while for steel work over $\frac{1}{4}$ in. the rightward method is used.

Before making the weld, preparation of the edges to be welded is necessary. This consists for the most part in **straightening up,** cleaning, and, on thicker plates above $\frac{5}{16}$ in., **vee-ing the edges.** For thin plates and **short runs** the **chamfers** to form the vee

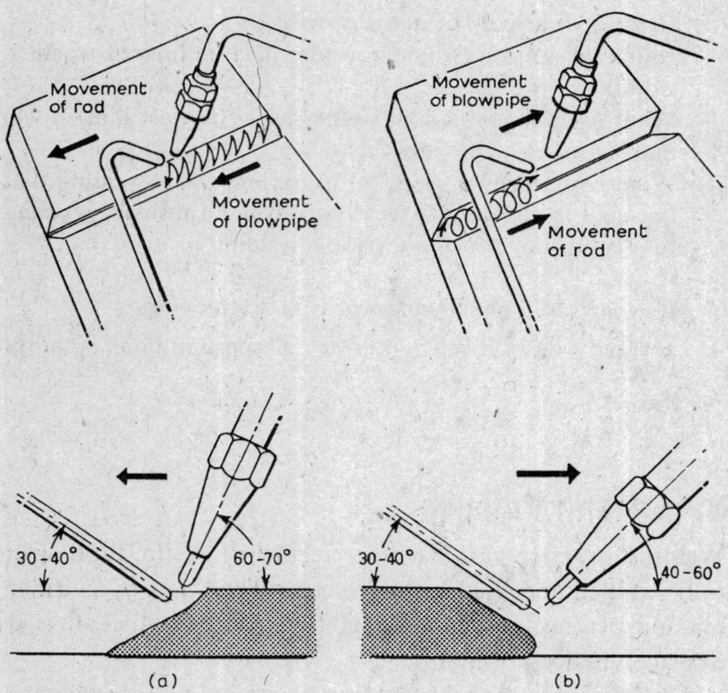

Figure 27 *Methods of Welding*
(a) the leftward (b) the rightward

can be filed or hand ground, but where possible they should be machined on a shaper, planer, or milling machine.

The welder must select a suitable filler rod and blowpipe tip for his work. These will vary in size according to the thickness of the metal to be welded. If the steel is of low carbon content the filler rod is of approximately the same composition and plated with copper to avoid **rusting.**

The actual **manipulation** of the blowpipe and filler rod, and the technique necessary for making a good weld, cannot be learnt from a book, but can only be acquired by practice under the **tuition** of a good welder. For this purpose, therefore, the reader should obtain **facilities** for practice, but there are several important points which we might impress upon him. Since welding is essentially a **casting process,** care must be taken that the joint is made by thoroughly fusing the metals together and not by sticking melted filler rod on to the surfaces to be joined. This will be avoided if care is taken to **ensure** that complete fusion penetrates over the entire surfaces to be welded. If the material being joined is veed for the purpose of adding additional metal from the filler rod, the aim should be to concentrate the heat of the flame into the vee until a molten pool of metal appears, and then add metal from the rod by dipping the end into the pool so that it melts and makes up the vee. With practice this can be performed as a progressive movement along the weld. For thick plates it may be necessary to fill in the bottom of the vee with one run and then repeat the process once or twice with additional layers of filling until the weld is completed. On no account should the work be hurried, but at the same time, particularly on thin plates, the flame should not be held too long in one place or a hole will be burnt in the plate. Thin plates are joined by melting their edges together without the addition of filling metal.

ARC WELDING PRACTICE

Arc welding practice differs from welding with oxy-acetylene equipment. You will remember that no blowpipe is used, but the welder is faced with the necessity of maintaining a regular arc. The preparation of the plates for welding is very similar to the method necessary for oxy-acetylene work and a suitable electrode must be selected according to plate thickness.

The electric current necessary to maintain the arc depends on the size and coating of the electrode. For bare or thinly coated rods the voltage varies from 15 volts on the small to 20 volts on the large diameter rods.

For medium or thickly coated rods the figures are 20–25 volts and 25–40 volts respectively. In order to complete the electric circuit with one pole connected to the electrode, the other lead must be attached to the work. With D.C. working, instructions as to whether the electrode forms the positive or negative pole is usually given with the electrodes; with A.C. operation the **polarity is immaterial,** since with alternating current the polarity reverses at the speed of the frequency.

When **commencing** a weld the electrode, fixed in the holder, is held in the right hand and the face shield in the left. The arc may be **struck** by **jabbing** the rod on to the plate and then withdrawing to about $\frac{1}{8}$–$\frac{1}{4}$ in. away, or by scraping the plate with it, using a slight circular motion so that the arc is struck at the bottom of the travel and further motion of the rod **elongates** the arc to the required length. When the arc is established the rod is held at a fairly steep angle and then moved slowly and evenly towards the operator. The deposited bead of metal should be continuous and even, with good penetration into the parent metal.

If the weld has been made with a medium or thickly coated electrode, when finished, it will be covered with a layer of slag which must be removed with pick, wire brush, or other convenient method.

Adapted from *Workshop Technology*, Part 3
by W. A. J. Chapman (Edward Arnold)

Explanation of words in bold type

$\frac{1}{32}$ in. upwards	$\frac{1}{32}$ in. or more in thickness.
leftward	going towards the left.
rightward	going towards the right.
straightening up	making straight.
veeing the edges	making the two edges form the shape of the letter V when placed side by side. (See figure 27.)
short runs	the 'run' is the length of the weld that the welder must move his blowpipe (or electrode) along.
chamfers	sloping edges.

rusting becoming corroded by oxidation. (See 'corrosion', Metals 5.)
manipulation handling, the way the welder handles (uses his hands to control) his equipment.
tuition teaching.
facilities are the things one needs to make work possible; in this case, the right time, place, and equipment for welding.
casting process a process in which the metals are melted.
ensure make sure (cf. **en**able – make able).
polarity is immaterial it does not matter which pole is which.
commencing beginning.
struck started.
jabbing touching with a quick, hard movement.
elongates lengthens.

Exercises

A Use these words to complete the following sentences: backward, chamfers, fused, leftward, manipulation, penetrate, plated, slag, strikes, vee.

1 The rightward method of making a butt weld may also be called the . . . . method.
2 For welding non-ferrous metals the . . . . method of working is considered to be the better.
3 When plates thicker than $\frac{5}{16}$ in. are to be welded it is necessary to . . . . the edges.
4 The edges which have been filed to form a V shape in the weld are called . . . .
5 To avoid rusting, filler rods composed of low-carbon steel are . . . . with copper.
6 A student cannot learn the correct . . . . of welding equipment simply by reading an instruction book.
7 Melted filler rod must not simply be stuck to the surface of the parent metals, since these must be completely . . . .
8 When beginning to weld with electric-arc equipment the welder . . . . an arc.
9 Some of the metal deposited from the filler rod must . . . . into the parent metal.
10 After welding with a thickly coated electrode . . . . must be removed from the surface of the weld.

B Answer these questions.

1 For what kind of butt welding is the rightward method used?
2 Describe a butt joint.
3 What must be done to prepare the plates for butt welding?
4 How can a student become a good welder?
5 Why must a welder avoid directing his flame at one point for a long time?
6 What instructions must a welder look for when he is going to work with direct current?
7 Give *two* ways in which a welder can strike an arc.
8 What does the welder have in his hands: (*a*) for oxy-acetylene welding; (*b*) for electric arc welding?
9 What distance from the work should an electrode be held after the arc has been struck?
10 Describe a correct vee weld.

C Use each of the following words in a sentence.

1. arc 2. permanently 3. elongate 4. coating
5. preparation

Concrete

1. Properties of Concrete

Concrete is produced by mixing together cement, water, and mineral aggregates. This mixture is placed into a suitable mould, **compacted,** and allowed to harden. It is somewhat similar to building stone but has the advantage that it can be easily moulded into any suitable shape, and also that it can be conveniently reinforced with steel rods to improve its structural properties.

A concrete mix may be regarded as being made up of two parts, first the aggregates (sand and stones) and second the cement paste (that is, water and cement). The cement paste covers the surface of the stones and sand particles, binding them together when it hardens. The aggregate is not altered in any way but is merely embedded firmly in a rock-like, hardened cement paste. This cementing material is formed by the water and cement combining together chemically by a process called 'hydration'.

The chemical reaction takes place quite slowly and continues for many years. The hardened cement paste becomes harder as the hydration continues and consequently concrete becomes harder and stronger as it grows older, provided that the temperature and moisture conditions are suitable. If, however, the concrete is allowed to dry out completely, its strength will increase no further. The longer it remains in a wet condition, the stronger it will become. It is also liable to attack by the weather or by chemicals, and it is necessary to ensure that it becomes strong enough to withstand such attacks.

When concrete is properly made and has been allowed to harden for a sufficient length of time (usually a few days), it can resist compression forces very well and can withstand quite severe weather conditions. It can resist **abrasion** and, although not absolutely waterproof, offers a relatively high resistance to moisture penetration.

One of its advantages is its **adaptability** for a wide range of uses. It is possible to make concrete of any required strength within reasonable limits, but the stronger it is made, the more it is likely to cost. It would be uneconomic to use concrete of a higher strength than is necessary for a particular job, but at the same time it must be strong enough to do its job properly. The concrete designer is thus able to use the type of concrete which will do its job efficiently for the lowest cost.

If we are to understand how to use concrete properly, it is important that we realise what its limitations are. From the structural designer's point of view, one of the main disadvantages of concrete is its low tensile strength. That is to say, it is not able to resist forces tending to pull it apart. This may be overcome by reinforcing with steel bars any part of a concrete structure where tensile stresses are likely to occur.

Concrete shrinks when it dries out and will expand and contract every time it is wetted and dried. This may set up tensile forces (or stresses) in the concrete and **shrinkage cracking** may occur. The designer can help to avoid this by specifying **contraction joints** in suitable places. The supervisor on the site can also help to reduce the effect of shrinkage if he ensures that the concrete is kept wet for as long as possible.

Another property of concrete which may be responsible for cracking is its expansion and contraction due to heating and cooling. The designer can overcome this by including **expansion** and contraction joints in suitable places, so as to allow the concrete to move freely when the temperature changes. Steel also expands when it is heated and it is very fortunate that it expands the same amount as concrete does for a given temperature rise. For this reason steel embedded in concrete will move with the concrete without setting up any tensile stress.

The efficient use of steel bars in concrete depends on the ability of concrete to grip the steel tightly enough to prevent it from pulling out. This interaction between steel and concrete is called 'bond', and the designer must check his calculations to ensure that the bond strength is sufficient. Bond may be reduced if the steel bars have loose scale, loose rust, or oil on the surface when they are embedded in the concrete. The supervisor should therefore check the steel to make sure that a good bond strength can develop.

It has already been mentioned that concrete is not entirely waterproof, but good concrete is quite waterproof enough for most practical purposes. It is, however, necessary to take most particular care with construction joints where there is a danger that such joints will allow water to pass through them.

Good-quality and poor-quality concrete might appear very similar **by visual inspection.** In order to assess the quality of the concrete properly it is necessary to use a more reliable method of inspection in the form of a suitable test. Experiments have shown that the test which generally gives the most reliable measurement of concrete quality is the crushing (or compressive) strength test. If a concrete has a high crushing strength it will usually have good durability (that is, resistance to weathering, chemical attack, and abrasion), as well as being strong enough to carry heavy structural loads. This test, which is convenient to carry out, is usually made on a cube which is crushed in a mechanical testing machine.

Adapted from *Concrete Practice*, Volume 1, by R. H. Elvery
(C. R. Books: Maclaren & Sons Ltd.)

Explanation of words in bold type

compacted compressed.
abrasion scraping.
its adaptability it can be used for many purposes.
shrinkage cracking cracks appearing because the concrete has contracted (grown smaller).
contraction joints joins made between concrete sections to allow each section to contract.
expansion joints joins which allow sections of concrete to expand.
by visual inspection when it is examined with the eyes.

Exercises

A Use these words to complete the following sentences: adaptable, binds, bond, expands, hardening, hydration, increases, steel, stronger, tensile.

1 Cement paste . . . . together the aggregates in concrete.
2 The chemical reaction set up between water and cement is called . . . .
3 Concrete is subject to the process of age . . . .
4 Concrete can be used for a great variety of purposes. It is an . . . . material.
5 The . . . . the cement, the more expensive it is.
6 The . . . . strength of concrete is low.
7 Reinforcing concrete with steel bars . . . . its tensile strength.
8 When concrete is wetted it . . . .
9 . . . . expands at the same rate as concrete when the temperature rises.
10 Oil on the surface of steel reinforcing bars will reduce . . . . strength.

B Answer these questions.

1 What materials are combined to make concrete?
2 What happens to the sand and stones in the process of making concrete?
3 How can concrete be made harder and stronger?
4 Give *four* reasons why concrete is a very useful building material.
5 What is one of the limitations of concrete?
6 What factors may cause cracking in concrete?
7 How can the tensile strength of concrete be increased?
8 Why must a supervisor carefully examine the steel bars used for reinforcing concrete?
9 What is the reason for using the following joints in concrete constructions: (*a*) contraction joints; (*b*) expansion joints?
10 Describe the test used to find out the crushing strength of concrete.

2. Reinforced Concrete

Mild steel has a high **tensile strength,** this being about 200 times that of concrete. It can be **embedded in** concrete so as to carry any tensile stress which may occur, such as in beams and slabs which are subjected to bending. Let us consider a plain concrete beam, that is, one having no steel embedded in it. If this beam is supported at its ends and loaded along its length it may break, as shown in figure 28. Such a failure would occur very suddenly and the beam would collapse completely.

You will notice in figure 28A that the underside of this beam is being pulled by tension stresses while the top side is being

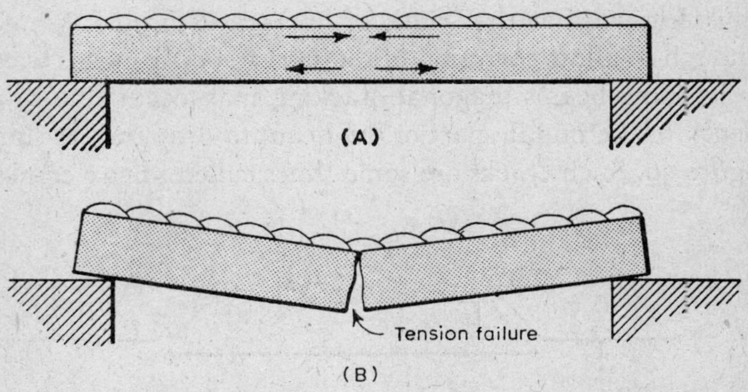

Figure 28 *Failure of a concrete beam with no reinforcement*

pushed by **compressive stresses.** The concrete is strong enough to carry the compression at the top, but is not strong enough to carry the tension at the bottom.

If a number of steel rods were embedded in the concrete near the bottom of the beam, the beam would be able to carry a much greater load before breaking, as the steel rods would have to be broken before the beam could collapse. The steel rods would not prevent the concrete from cracking, and, in fact, most reinforced concrete beams are cracked when they are carrying the load for which they were designed. This is illustrated in figure 29.

The cracks so formed need not worry us since the steel rods would be quite capable of holding the two halves together safely.

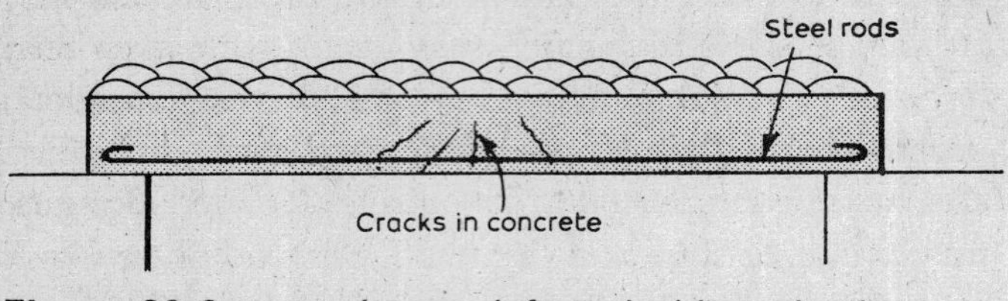

Figure 29 *Concrete beam reinforced with steel rods*

There may, however, be a danger that moisture would enter the cracks and cause the steel to rust. The designer of the beam must ensure that the width of any crack formed is not large enough to allow corrosion of the steel to occur in this way.

In some beams diagonal cracking may occur due to a tendency of the middle part of the beam to drop out, as shown in figure 30. Such cracks are some times called **shear** cracks, and

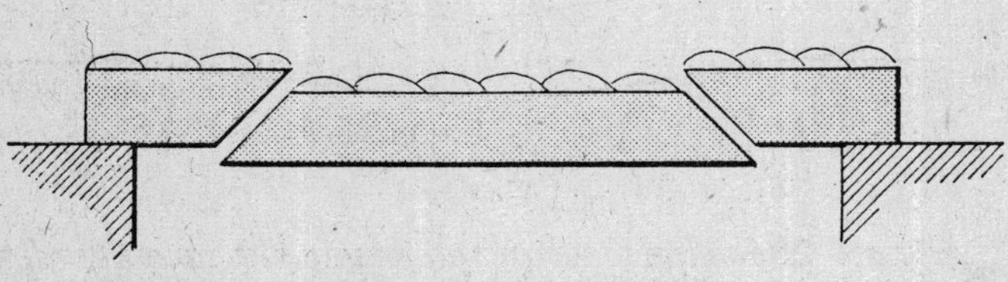

Figure 30 *Failure of a beam by shear cracking*

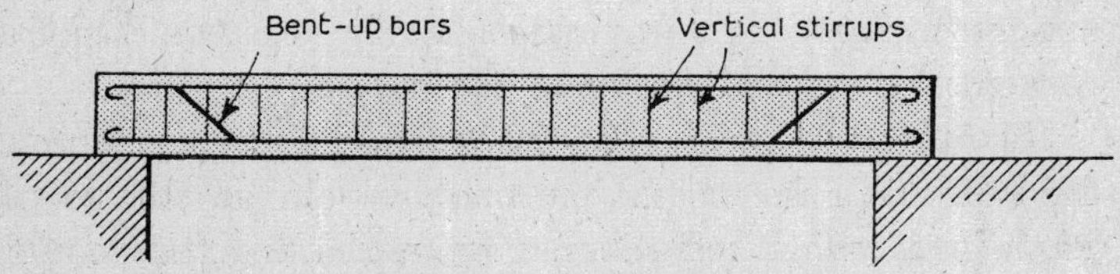

Figure 31 *Beam reinforced to resist shear cracking*

to avoid this, the designer may specify the inclusion of vertical stirrups or bent-up bars similar to those illustrated in figure 31.

The best position of the steel in a beam will depend on how the beam is supported as well as on the loading it is expected to carry. For example, **a cantilevering beam** bends so that its top surface is in tension and its bottom surface is in compression. The main reinforcement must therefore be embedded near the

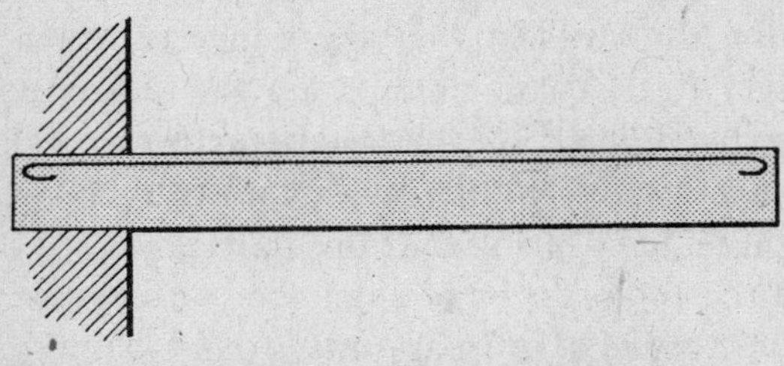

Figure 32 *A reinforced cantilevering beam*

top surface of the beam (figure 32) and cracks may occur near the support at the top.

Generally the design of reinforced concrete is the responsibility of the engineer, who sends instructions and drawings to the supervisor on the site. If these instructions are carried out exactly as required by the engineer it is not really essential for the **site personnel** to understand the reasons for the instructions. On the other hand, if the site supervisor has a clear understanding of the elementary principles involved in reinforced concrete design, he can interpret the drawings and instructions more intelligently and should find it easier to co-operate with the engineer. This is particularly so in appreciating the need for accuracy in positioning steel reinforcement.

Let us consider the steel in a reinforced concrete beam. If the steel is placed too near the outside surface of the beam, the thickness of the concrete covering the steel may become too small. It has been found by experience that a certain minimum amount of cover is necessary so that the steel is protected from moisture and prevented from rusting. If the concrete cover is too small the steel will rust and this rust will push the concrete cover off, exposing the steel to even more severe corrosion. We must not, however, place the steel too far from the surface, since this may have the effect of weakening the beam in another way. It can be seen from figure 29 that steel has its greatest effect in strengthening the beam if it is placed as near to the bottom surface as possible. If we were to place the steel half-way between the top and bottom surfaces of the beam it would not help to strengthen the beam very much. The engineer's drawings will show the best position for the steel so that the concrete cover is sufficient to protect it. His calculations for the strength of the beam will have been based on the **assumption** that the steel will be placed accurately, and he will expect the co-operation of the site personnel in ensuring that this is done.

There are three main types of steel used in concrete. The commonest type consists of ordinary mild-steel bars which vary in diameter from $\frac{3}{16}$ in. to 2 in., although bars greater than 1 in. diameter are only used in large-scale construction work.

High-tensile steel bars are also used for ordinary reinforced concrete and are of similar diameter to mild-steel bars. This type of steel is normally allowed to carry one and a half times the load which mild steel of the same diameter can carry.

High-tensile wires and alloy-steel bars used entirely for prestressed concrete, which will be dealt with in the next passage, have much higher strengths than the steels used for ordinary reinforced concrete. The **allowable stresses** used for these high-tensile wires may be up to about eight times that for mild steel and the stress for the high-tensile alloy bars is about four and a half times the mild-steel stress.

Steels for ordinary reinforced concrete may be obtained in a variety of forms apart from the normal round bars. These include deformed and **indented** bars and a variety of various types of twisted steel. These bars can usually be relied upon to provide a better bond with the concrete than that obtained with ordinary round bars. In addition to the improved bond, some of these types have a higher tensile strength and can be allowed to carry a greater load.

Mild steel has a high **ductility** compared with most of the high-strength steels, and this enables it to stretch without breaking when it is suddenly overloaded beyond its ordinary safe working load. This also enables it to be bent readily into the required shapes.

The type of high-tensile steel used for ordinary reinforced concrete is less ductile than mild steel and is more difficult to bend, but it is generally quite suitable for bending into the shapes normally required.

Adapted from *Concrete Practice*, Volume 1 by R. H. Elvery (C. R. Books: Maclaren & Sons Ltd.)

Explanation of words in bold type

tensile strength strength to resist forces tending to pull materials apart.
embedded in fitted into, enveloped in.
compressive stresses stresses which are forcing the material to press together.
to shear to cut.
a cantilevering beam projects from the side of a wall or pillar to support part of a structure, a balcony, for example.
site personnel persons working on the site.
assumption belief.
high-tensile steel bars bars with high tensile strength.
allowable stresses the stresses that it is allowed, or safe, to put on these wires.
indented bars have dents, or 'notches', cut into their surface.
ductility the quality of not breaking when stretched (adjective – ductile).

Exercises

A Use these words to complete the following sentences:
bond, bottom, cantilevering, compression, corroded, ductility, greater, load, rust, wires.

1 Mild steel has a much . . . . tensile strength than concrete.
2 Concrete can withstand . . . . better than it can withstand tension.

3 In most reinforced concrete beams steel rods are placed near the . . . . surface.
4 In . . . . beams reinforcement is necessary near the top surface.
5 Reinforcing steel rods may be . . . . if cracks allow moisture to reach them.
6 . . . . from reinforcing rods may push up the concrete surface covering them.
7 High-tensile steel bars can be allowed to carry a greater . . . . than ordinary mild-steel bars.
8 Indented bars form a better . . . . with the enveloping concrete than round bars.
9 High-tensile . . . . can be subjected to a stress of about eight times that for mild steel.
10 Mild steel can be stretched without being broken because it has a high . . . .

B Answer these questions.

1 Why are the steel rods embedded near the bottom of the beam in figure 29?
2 How may beams be reinforced to avoid shear cracks?
3 Where does tension occur in a cantilevering beam?
4 How does a site supervisor know where to place the reinforcement in a beam?
5 Why must reinforcing rods not be placed too near the surface of a beam?
6 What are the three main types of steel used in reinforced concrete?
7 What is the advantage in using indented steel bars and twisted steel instead of round bars in reinforcing concrete?
8 Give *two* reasons why mild steel is a particularly useful material for reinforcing concrete.
9 Describe the reinforced concrete beam illustrated in figure 31
10 Describe a reinforced concrete cantilevering beam.

C Use each of the following words in a sentence.

1. enveloped 2. stresses 3. waterproof 4. reinforce 5. ductile

3. Prestressed Concrete

Mild-steel bars are used in ordinary reinforced concrete beams to overcome the tensile weakness of concrete. Another way of overcoming the weakness is by **prestressing** the concrete so as to induce in it compression stresses before it is subject to the normal tensile stresses due to bending. When the beam is loaded and bending occurs, **the bending tensile stress is neutralised** by the prestressing compression so that no actual tensile stress develops. In this way the concrete is used more

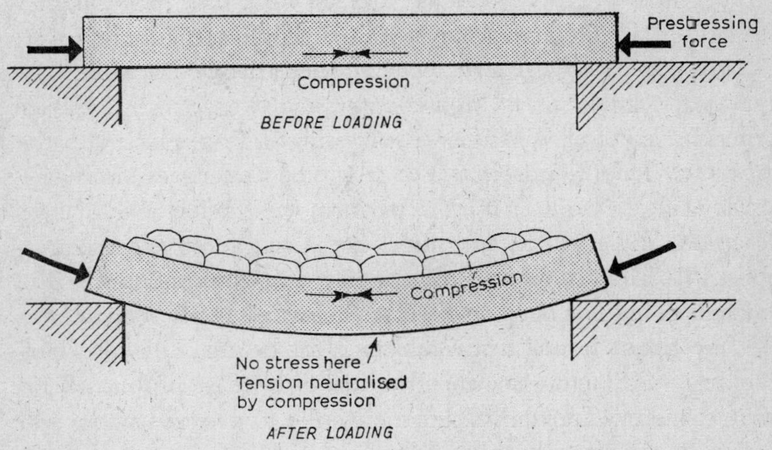

Figure 33 *Principle of a prestressed concrete beam*

efficiently and no tensile cracks occur as they do in ordinary reinforced concrete. This principle is illustrated in figure 33.

The prestressing may be done in several different ways, and these are discussed later. Usually the prestress is achieved by using high-tensile steel wires or alloy-steel bars. This steel is about two and a half to five times as strong as ordinary mild steel. Generally less concrete and much less steel is required for a prestressed concrete beam compared with a reinforced concrete beam of the same strength. This means that, for prestressed concrete, the weight of the structure will be less than if it were built of ordinary reinforced concrete. It is necessary, however, for the quality of the concrete to be much higher than that used for ordinary concrete construction and special techniques are required for the prestressing operations.

Prestressed concrete structural **members** are usually more springy than ordinary reinforced concrete, and behave rather more like timber in having a high degree of **resilience.** In some circumstances the technique is only economic when used in the form of factory-built **precast units,** but in many cases the prestressing can be applied successfully to **in-situ** concrete, particularly where large **spans** are concerned, and where the transporting of large units is difficult or impossible.

The concrete required for prestressed work must have a higher strength than that used for ordinary reinforced concrete, but this has the advantage that the concrete is **more durable.** The concrete is subjected to high compressive stresses and these are generally greatest when the prestressing is first applied to the concrete. This means that the concrete undergoes a kind of test loading before it is used to perform the job for which it was designed, and if it can withstand the initial prestressing conditions, it is likely to be well able to carry its **design load** safely.

Prestressing has been applied to almost all types of structure, but **the most usual application** is for beams. It is also particularly suitable in the construction of precast pipes and for large circular water tanks, since it is able to provide a concrete wall quite free from cracks.

There are two distinctly different methods of prestressing, one suitable for factory products and the other for use **in the field.** Both methods use high-tensile steel to apply prestress to the concrete, but in one case the steel is tensioned before the concrete hardens (pre-tensioned), and in the other it is tensioned after the concrete hardens (post-tensioned).

Pre-tensioning is suitable for factory-produced units and is usually done by a 'long line' process. A series of moulds is laid in a long line and a single group of wires is stretched from one end of the line to the other, so that it passes through the ends of all the moulds. Concrete is then placed into the moulds around the stretched wires and, after compacting, is allowed to harden. When the concrete has reached the necessary strength the moulds are removed and the wires **protruding** from the ends of the concrete units are cut off level with the concrete. When they have been cut, the wires become shorter as they attempt to return to their original unstretched length. They are not able to shorten to their original length, since they are firmly

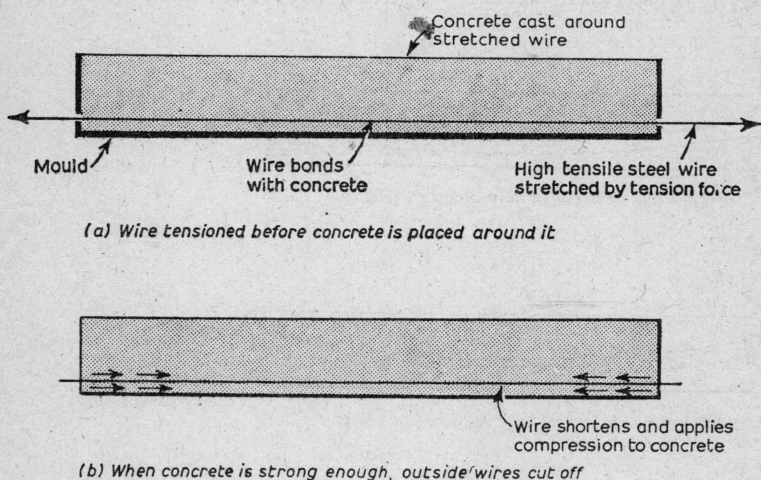

Figure 34 *Pre-tensioning*

gripped by the concrete, but they do shorten slightly and in doing so exert a compressive stress on the concrete. This is illustrated in figure 34.

Post-tensioning techniques are of more direct interest to those engaged on site work, since it may be necessary to apply some of these techniques in practice.

Several systems are in use for applying tension to the steel and for providing anchorages to maintain the tension permanently. Whichever system is used, the steel is required to slide freely inside the concrete when the prestress is being applied. For this reason post-tensioning is often referred to as a 'non-bonded' method.

In order to prevent the steel from bonding with the concrete during the prestressing process, the steel wires or rods are either inserted into **preformed ducts** in the concrete after this has hardened or are laid in the **formwork** with suitable **sheathing** around them so that when the concrete is placed it will not come into contact with the steel.

The cable is anchored at one end and a **jack** is applied to the other end. This jack grips the cable and pulls it with the required prestressing force, while at the same time pushing

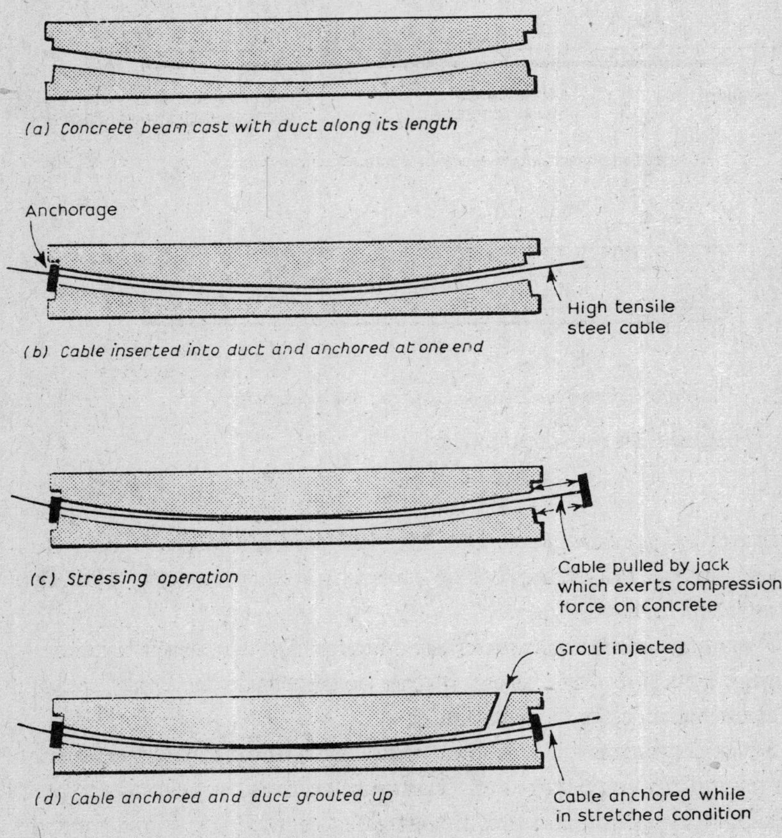

Figure 35 *Post-tensioning*

against the end of the concrete. While the cable is in the stretched condition, its end is anchored permanently to the concrete and the jack is removed. The final operation is the injection of **cement grout** in to the cable duct to protect the cable from any possible corrosion. The operations involved in post-tensioning are illustrated in figure 35.

The prestressing operation must be done carefully and according to the engineer's instructions. These instructions will include the amount of force to be exerted on the cable and also the amount by which it must be stretched.

When the cable is very long or when it is placed in a duct with sharp bends the stressing is done from both ends by two jacks operated **simultaneously.** The amount of stretch required for a long cable may be greater than the extending capacity of a single jack, so that two are needed.

Whenever the stressing is being done the area behind the jack should be regarded as a danger zone to be kept clear in case the cable breaks suddenly. A considerable amount of energy is stored up in the cables when they are prestressed, and if they break, this energy is released suddenly and violently.

Adapted from *Concrete Practice*, Volume 1 by R. H. Elvery
(C. R. Books: Maclaren & Sons Ltd.)

Explanation of words in bold type

prestressing compressing the concrete before it is used (cf. **pre**viously in Petroleum 2).

the bending tensile stress is neutralised the effect of this stress is taken away (or 'cancelled out').

members beams, pillars, arches, etc.

resilience the ability to bend under pressure without breaking.

precast units sections of a structure cast, or moulded, before they are taken to the site (cf. **pre**stressed).

in-situ a Latin phrase meaning 'on the site'.

spans are horizontal members of a construction supported from below at each end.

more durable will last longer.

design load the load which it is designed to carry. The 'load' is the amount of force pressing down on to a concrete beam.
the most usual application what it is most often used for.
in the field on the site.
protruding projecting (sticking out).
preformed ducts 1. ducts – see Ventilation 4.
2. preformed – formed before the concrete is allowed to harden (cf. **pre**stressed, **pre**cast).
formwork the frame of the mould into which the concrete is put. Formwork may be made from timber or steel.
sheathing a coating.
jack a machine for lifting or pulling something.
cement grout a fluid mixture of cement.
simultaneously at the same time.

Exercises

A Use these words to complete the following sentences:
after, corrosion, bonded, jack, less, neutralises, precast, prestressing, stretched, stronger.

1 Prestressing of concrete . . . . the effect of the tensile stress that a beam may be subjected to.
2 Prestressed concrete members weigh . . . . than ordinary reinforced concrete members of the same strength.
3 The members may be . . . . or formed on the site.
4 Concrete used for prestressed work must be . . . . than ordinary reinforced concrete.
5 The greatest compressive stresses to which prestressed concrete beams are subjected are usually applied during the . . . . process.
6 In the post-tensioning method stress is applied to concrete . . . . it has hardened.
7 Wires which have already been . . . . are embedded in the concrete in the pre-tensioning method.
8 The steel wires are not . . . . in the post-tensioning method.
9 The wires are shielded from . . . . by cement grout.
10 The cable passing through the concrete is stretched by a . . . .

B Answer these questions.

1 What kind of steel is used for: (*a*) ordinary reinforced concrete; (*b*) prestressed concrete?
2 Why will the weight of a structure built of prestressed concrete be less than one built of ordinary reinforced concrete?
3 Name three types of structural members which may be made of concrete.
4 In what circumstances may prestressed concrete be formed on the building site?
5 Why is prestressed concrete particularly suitable for constructing water tanks?
6 How do the wires used in the pre-tensioning method exert a compressive stress on the concrete?
7 Why is the post-tensioning method often referred to as a 'non-bonded' method?
8 Why should site personnel avoid passing behind the jack while concrete is being prestressed?
9 What are the two functions of the jack while concrete is being prestressed by the post-tensioning method?
10 Describe the concrete beam illustrated in figure 35 (*d*).

C Use each of these words in a sentence.

1. embedded 2. durable 3. load 4. age hardening 5. precast

Index

Numbers refer to pages on which words and phrases are first explained